Computational Intelligence Methods and Applications

The monographs and textbooks in this series explain methods developed in computational intelligence (including evolutionary computing, neural networks, and fuzzy systems), soft computing, statistics, and artificial intelligence, and their applications in domains such as heuristics and optimization; bioinformatics, computational biology, and biomedical engineering; image and signal processing, VLSI, and embedded system design; network design; process engineering; social networking; and data mining.

More information about this series at http://www.springer.com/series/15197

Marcin Relich

Decision Support for Product Development

Using Computational Intelligence for Information Acquisition in Enterprise Databases

Marcin Relich
Faculty of Economics and Management
University of Zielona Góra
Zielona Góra, Poland

ISSN 2510-1765 ISSN 2510-1773 (electronic)
Computational Intelligence Methods and Applications
ISBN 978-3-030-43899-9 ISBN 978-3-030-43897-5 (eBook)
https://doi.org/10.1007/978-3-030-43897-5

This Springer imprint is published by the registered company Springer Nature Switzerland AG.
The registered company address is: Gewerbestrasse 11, 6330 Cham, Switzerland

Preface

New product development (NPD) is a complex, long-lasting, and risky process affecting an enterprise's competitiveness and survival. As many product development efforts are futile, an enterprise usually develops simultaneously a few products to increase the chance of launching a successful product on the market. Consequently, there is a need to select NPD projects with great potential and manage them effectively. There are several methods for evaluating the potential of a new product, embracing financial and non-financial criteria (e.g. widely used net present value (NPV) and scoring methods). These methods allow decision-makers to obtain information about the attractiveness of product concepts, enabling project ranking and identification of an optimal NPD portfolio. If an NPD project does not fulfil acceptance criteria, then it is not further considered. However, the decision-maker may be interested in obtaining information about prerequisites by which a specific NPD project could meet selection criteria. This information seems to be particularly important in the case of strategic NPD projects that aim to improve a company's competitiveness. Moreover, decision-makers' expectations refer to an adaptable NPD model that can be easily updated according to new conditions. A project-enterprise model should enable effortless specification of NPD-related problems and the answer to questions: Are there all appropriate resources to add a new project to the current NPD portfolio? What changes should be introduced to optimise the NPD portfolio? Nowadays, there is a lack of methods dedicated to the above-mentioned expectations. This provides the motivation for elaborating an approach that supports the decision-maker in evaluating product concepts, selecting the project portfolio, and identifying prerequisites for achieving a desirable outcome of an NPD project.

This study aims to examine the thesis that some techniques for acquiring information and specifying an NPD model may be successfully used in conducting dependency analysis of business components and simulations of economic effects of managerial decisions. These techniques embrace problem specification in a declarative representation and the use of computational intelligence for identifying relationships, applying them to evaluate the product's potential, and seeking possible variants of the NPD project implementation that fulfil the decision-maker's requirements. The intention of this monograph is an attempt to answer the following questions:

- What should a methodology of creating an NPD model be to ensure reliable mapping of decision problems that occur in the NPD process?
- How should computational intelligence techniques be adjusted for evaluating the potential of a new product, especially within parametric estimation?
- How should a decision support system be constructed to enable effective analysis of economic effects of possible managerial decisions within the portfolio selection and resource allocation problem?

The proposed NPD model consists of variables and constraints related to a new product, enterprise, and its environment. Constraints can refer to available resources, dependencies between NPD projects, and relationships between input and output variables. The NPD model is a platform for formulating NPD-related problems such as evaluation of the potential of a new product, selection of a portfolio of NPD projects, and scheduling them. As the NPD model includes cause-and-effect relationships between specific input and output variables, the mentioned problems can be stated in a *forward* or *inverse* form. The solution of the problem stated in the forward form is related to identification of an outcome (e.g. the NPV from a new product) on the basis of the values of input variables. In turn, the problem stated in the inverse form is solved through seeking possible values of input variables to reach a desirable outcome. If there is no satisfactory solution of the problem stated in the forward form, then the problem is reformulated into the inverse form.

The solution of the problem stated in the forward and inverse form may be referred to prediction and simulation, respectively. Using relationships between input and output variables, the value of an output variable is predicted or possible variants of alternative performance of an NPD project are simulated. The problem stated in the inverse form allows the decision-maker to obtain information about the possibility of reaching a desirable outcome. The proposed approach enables the identification of all possible solutions (if they exist) of a problem taking into account dependencies between variables and other (e.g. resource-related) constraints. As a result, the proposed approach is able to find a wider range of simulations for economic effects of product development in comparison with the traditional scenario analysis.

The NPD model is specified in terms of a constraint satisfaction problem (CSP) that consists of variables, their domains, and constraints that link these variables and reduce the number of solutions. The use of the CSP paradigm enables formulation of the NPD model in a declarative representation, in which the desired results are specified without explaining the specific algorithms needed to achieve these results. Current NPD models are based on a procedural approach that uses a set of defined algorithms to solve a specific problem. Specification of the NPD model in terms of the CSP enables consideration of several NPD-related problems within a single NPD model. Moreover, the CSP makes it possible to use the problem specification in the forward and inverse form alternately.

The model formulated in a declarative representation enables its immediate adaptation to new conditions related to the changeable business environment (e.g. customers' needs, competitors' responses, accessible resources). Consequently, decision support systems (DSSs) designed according to a declarative approach tend to a fuller presentation of a decisive situation. Current DSSs mainly use a procedural approach, in which predetermined algorithms are designed for solving a problem, limiting adaptable properties of DSSs. The NPD model specified in terms of variables and constraints also enables problem statement in the inverse form. As the search space depends on the number of decision variables and their domains, there is usually an enormous number of admissible solutions. Effective identification of all admissible solutions of NPD-related problems is then a challenging task. This is an incentive to use constraint programming (CP) to reduce the search space and solve the problem formulated in terms of the CSP. The application of CP enables time reduction of finding solutions and consequently improves interactive properties of DSSs.

The proposed approach is dedicated to project-oriented enterprises that develop new or improved products and store the data related to previous NPD projects in enterprise databases. Improved products are often introduced by enterprises belonging to industries such as electronics (e.g. computers, smartphones, TVs), automotive (e.g. parts related to a new car's design), and white goods (e.g. fridges, blenders, hair dryers). Enterprise databases can be analysed towards acquiring information for supporting decision-makers in evaluating the attractiveness of a new product and identifying potential difficulties during project execution. Information acquisition requires specific techniques to cope with selection of the most relevant input variables, identification of relationships between input and output variables, and prediction. The need to discover non-linear relationships in multidimensional datasets and incorporate imprecision in evaluating product development leads to the use of computational intelligence (CI) techniques (namely artificial neural networks and neuro-fuzzy structures). CI techniques are used to solve problems of information processing that are ineffective or unfeasible when solved with traditional approaches based on statistical modelling. Moreover, the heuristic nature of CI techniques facilitates their use in the product development environment that often involves factors specified in an imprecise form.

To sum up, the proposed approach includes the following characteristics:

- The NPD model formulated in a declarative representation facilitates model updating and enables the formulation of NPD-related problems in the forward and inverse form.
- The use of CI techniques facilitates information acquisition from enterprise databases, including pattern recognition within multidimensional data structures and the imprecise nature of product development data.
- The use of CP reduces the search space and time for finding admissible solutions and improves interactive properties of DSSs.

The monograph consists of six chapters. Chapter 1 presents the state of the art and challenges in product development. This chapter is concerned with describing product development in terms of a systems approach, product development phases, and portfolio management, including current methods for evaluating the potential of a new product and selecting a portfolio of NPD projects. A literature review indicates the importance of extending current research towards designing an NPD model within a declarative approach.

Chapter 2 is concerned with elaborating a methodology to build the NPD model that ensures reliable mapping of decision problems within product development. The proposed model is dedicated to three problems: evaluation of the product's potential, portfolio selection, and resource allocation between ongoing and new projects. The model of evaluating the product's potential includes variables embracing product characteristics (e.g. its size, weight, number of parts), enterprise (e.g. the cost of labour, product price), and its environment (e.g. the cost of materials, sales volume). The model of a project portfolio includes ongoing NPD projects and new projects that are considered for the portfolio. A new project is added to the portfolio if its evaluation indicates the desirable potential. Adding a new project to the portfolio or removing an ongoing project from the portfolio requires scheduling according to the priority of an NPD project. Moreover, this chapter provides foundations to formulate NPD-related problems in terms of the CSP in order to facilitate adaptation of a model to new conditions and enable the problem statement in the forward and inverse form.

Chapter 3 is concerned with designing a method for an effective solution of NPD-related problems. The proposed method is based on the application of some computational intelligence and constraint programming techniques to improve problem-solving. This chapter proposes a methodology for acquiring information from enterprise databases and evaluating the potential of a new product in terms of parametric estimation. The relationships between variables are identified with the used of artificial neural networks (ANNs) and compared with multiple regression. The presented results indicate that ANNs are able to estimate the output variable more precisely than multiple regression, allowing decision-makers to obtain reliable information from enterprise databases in order to identify the potential of a new product. Moreover, a neuro-fuzzy system is used to take into consideration customers' opinions about a new product specified in an imprecise form. Identified cause-and-effect relationships are further used to carry out simulations within NPD-related problems stated in an inverse problem. The performed experiments show that CP significantly reduces the search space and time needed to obtain results in comparison with the entire search space.

Chapter 4 presents the foundations of designing a decision support system in terms of a declarative approach. The proposed DSS includes dependency analysis between input and output variables, prediction of an output variable (e.g. the NPD cost), portfolio selection of NPD projects, and optimal resource allocation for a project portfolio. Chapter 4 also verifies the application of the proposed approach in project-oriented enterprises.

Chapter 5 presents managerial implications for implementing a declarative approach and CI techniques in project-oriented enterprises. This chapter aims to present advantages and limitations of the proposed approach in the context of the widely used NPV method. The presented examples compare the traditional scenario and sensitivity analysis with the proposed approach. Chapter 6 summarises the main findings of the monograph. I would like to acknowledge all the individuals and institutions that have supported me during the work on this book. I would like to especially thank my mentor Professor Zbigniew Banaszak, Koszalin University of Technology, for his helpful comments and support. I am also extremely grateful to my family—Joanna and Michał, who accepted the temporary loss of husband and father during the preparation of this book.

Zielona Góra, Poland
February 2019

Marcin Relich

Contents

Chapter 1
Product Development: State of the Art and Challenges

Product Development in Terms of a Systems Approach

The aim of a company is to obtain the profit through providing products or services to customers. Products are developed and manufactured with the use of the company's resources such as people, money, and machines. Product development also depends on the company's environment, including customers' needs, materials and technologies provided by suppliers, substitutional products provided by competitors, and legal regulations related to environmental and safety requirements. Figure 1.1 presents the company, its environment, and products in terms of a systems approach. The company's resources can be treated as an input, whereas products as an output of the system.

The systems approach may be defined as a logical process of problem-solving (Kerzner 2001). The word *process* indicates an ongoing system that is fed by input from its parts. The systems approach requires review of interrelationships between the various subsystems, aiming to seek an optimal solution or strategy in solving a problem. One of the fundamental characteristic of the systems approach is objective thinking, in which alternatives are determined through viewing events, phenomena, and ideas as external and apart from self-consciousness of potential decision-makers. Consequently, there is a possibility of recognising the existence of alternatives that could be omitted through personal bias in individuals.

A framework of the systems approach is presented in Fig. 1.2. Objective is the function of the system or the strategy that must be achieved. Requirements refer to partial needs to satisfy the objective. Alternatives are seen as the selected ways to implement and satisfy requirements. The most preferable alternative is determined according to selection criteria. Constraints are related to accessible resources and requirements that the alternatives must meet. Finally, the decision-maker obtains a proposal of solving a problem (e.g. selection of project portfolio). The mentioned

M. Relich, *Decision Support for Product Development*, Computational Intelligence Methods and Applications,
https://doi.org/10.1007/978-3-030-43897-5_1

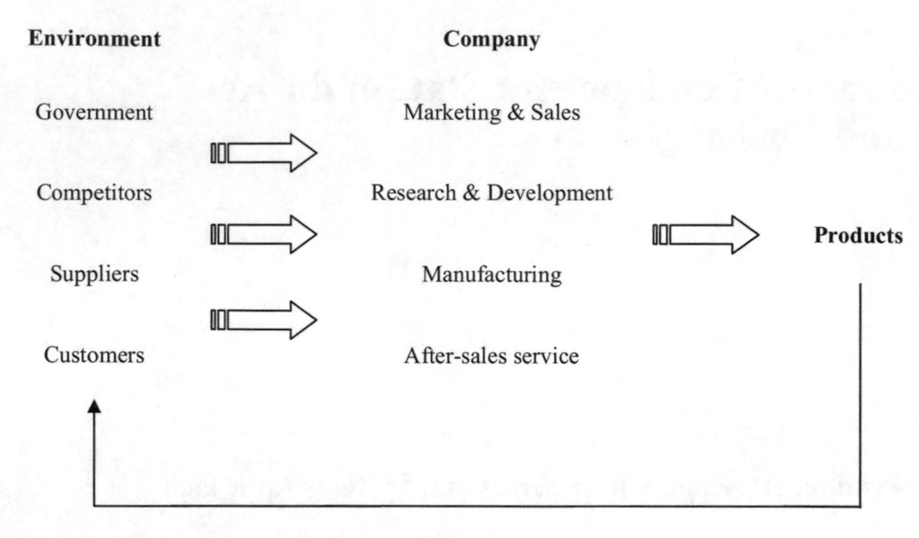

Fig. 1.1 Company in terms of a systems approach

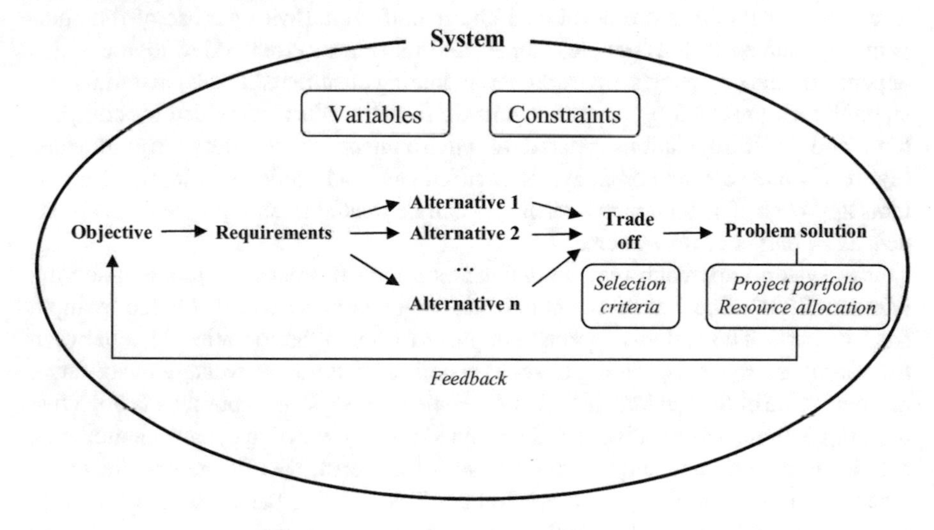

Fig. 1.2 The systems approach

parts of the systems approach may be described with the use of variables and constraints.

Figure 1.3 presents the systems approach for solving the problem of selecting a portfolio of NPD projects with the most potential. Feedback between output and input of the system can be considered as using past experiences to improve new product development. Measuring results and comparing them with predictions support learning capabilities of the system.

The systems approach to problem-solving can be divided into the following phases: problem formulation, analysis, and synthesis. In the first phase, problem

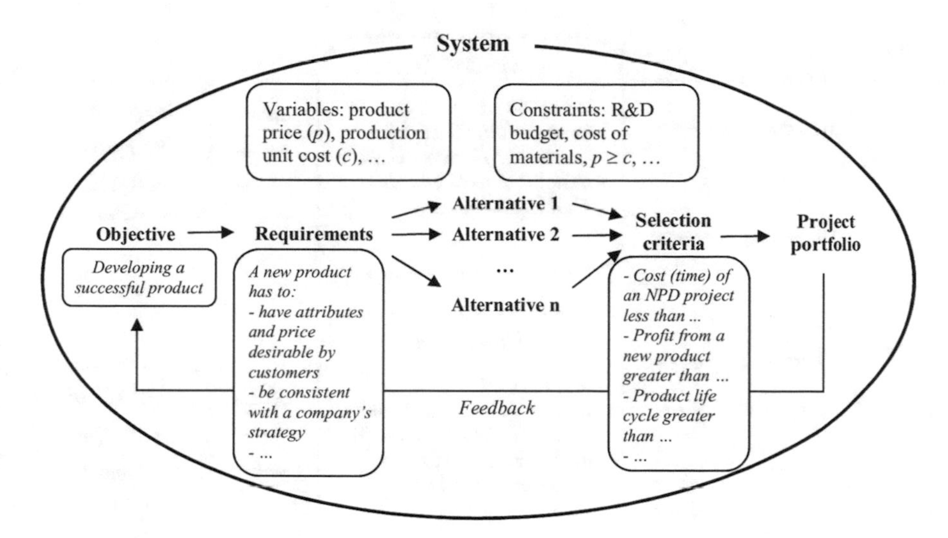

Fig. 1.3 The systems approach for selecting project portfolio

objective, variables, and constraints are defined in terminology accepted by project participants. The phase of analysis includes the identification of possible alternatives to the solution of the problem. In turn, the phase of synthesis is concerned with reaching the objective of the system through selecting the best solution.

The systems approach presented in Fig. 1.3 allows the decision-maker to obtain information about predicted values of selection criteria for all alternatives that correspond to requirements. The more requirements are considered, the more alternatives are generated. However, there is no guaranty that project portfolio will be reached. If any alternative cannot fulfil selection criteria, project portfolio will be empty or unchanged. Then there are sought prerequisites that fulfil selection criteria and reach a portfolio of NPD projects. Figure 1.4 presents two perspectives of problem statement: for a question stated in a *forward* form ('what profit is related to an NPD project for given prerequisites?'), and for a question in an *inverse* form ('what prerequisites must exist to ensure a desirable profit?'). Figure 1.4 illustrates these two perspectives with reference to a single criterion of portfolio selection—the profit from a new product.

The use of the proposed approach allows the decision-maker to identify prerequisites that enable product development with the desirable profit and according to the specified constraints. The number of possible variants of NPD project performance depends not only on assumed constraints but also on the precision of decision variables (e.g. the cost of product development, marketing cost, unit production cost, sales volume, product life cycle). The data related to these variables can be retrieved from enterprise databases and used to identify cause-and-effect relationships between input and output variables in order to predict the product's potential (the problem stated in a forward form) and search possible changes for reaching the expected value of an outcome (the problem stated in an inverse form). Figure 1.5

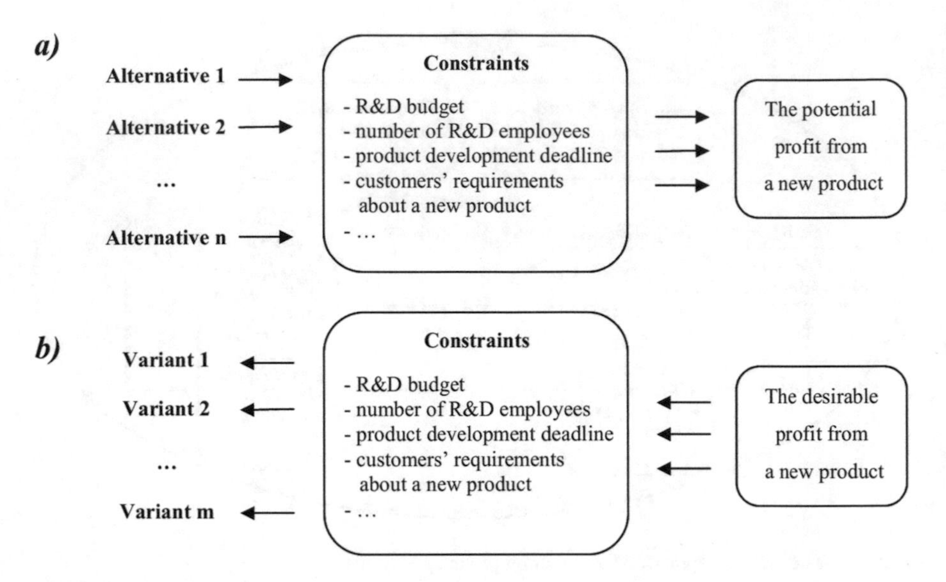

Fig. 1.4 Problem statement in a forward (**a**) and inverse form (**b**)

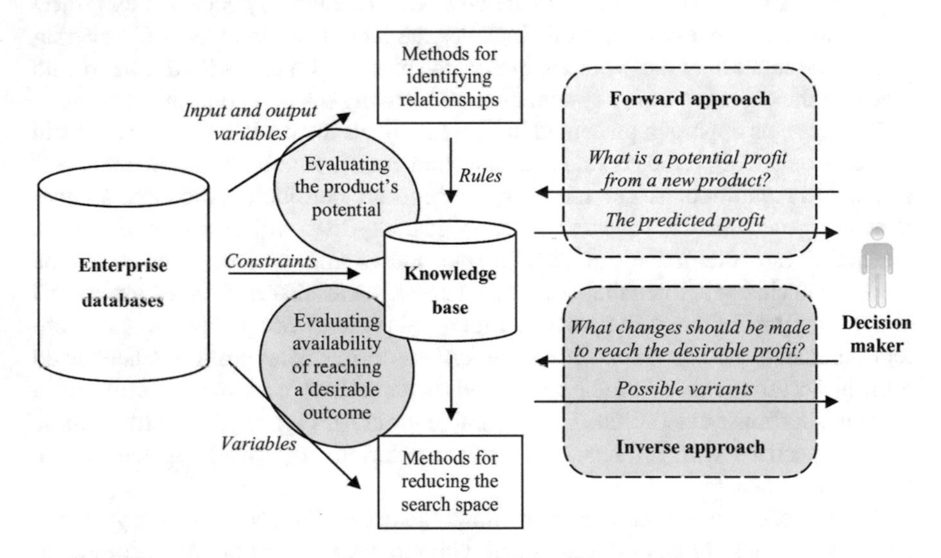

Fig. 1.5 The proposed knowledge-based system for solving the problem stated in a forward and inverse form

presents a framework of a knowledge-based system for solving NPD-related problems stated in a forward and inverse form.

An enterprise system includes software packages that collect data from the company and its environment, in order to support business processes, information flows, reporting, and data analytics. An example of this system is software packages regarding enterprise resource planning (ERP), customer relationship management (CRM), or computer-aided design (CAD). As a result, enterprise databases can provide facts and constraints to a knowledge base. Moreover, enterprise databases are a potential source of important information for evaluating the potential of a new product. However, information acquisition from large databases is a non-trivial process, in which a dedicated method should be developed towards identifying the most significant input variables and relationships among data related to previous NPD projects and predicting the value of an output variable. For this purpose, the use of some computational intelligence techniques has been proposed and verified.

The proposed knowledge-based system supports the decision-maker in answering three standard questions:

1. What is the value of an output variable (e.g. the profit from a new product)? If this value is unsatisfactory for the decision-maker, then this question is reformulated towards an inverse form as follows:
2. What changes in decision variable(s) should be made to obtain the expected value of an output variable? If there is no solution or it does not satisfy the decision-maker, then this question is reformulated towards seeking changes in constraints (e.g. the deadline or budget of an NPD project) to the following form:
3. What changes in constraints should be made to obtain the expected value of an output variable?

The second and third question refers to the problem stated in the inverse form, and finding possible solutions to this problem is related to search in a potentially vast space. Hence, there is a need to develop a method for reducing the search space in the context of solving the problem stated in the inverse form. In this study, constraint programming has been proposed to implement and solve problems formulated in terms of a CSP.

The proposed approach includes several advantages such as the specification of possible alternatives for product development (for questions stated in a forward and inverse form), the improvement of forecasting accuracy compared to multiple regression analysis, and the easiness of adding constraints to an NPD model and developing a decision support system. The comparison of the traditional and proposed approach is presented in Table 1.1.

Table 1.1 The comparison of the traditional and proposed approach

Traditional approach	Proposed approach
The evaluation of alternatives in terms of a potential outcome (problems stated in a forward form)	The evaluation of alternatives in terms of a potential outcome, and prerequisites that should be met to reach the desirable value of an outcome (problems stated in a forward and inverse form)
The outcome is predicted by analogical reasoning or parametric estimation based on multiple regression	The outcome is predicted by parametric estimation based on computational intelligence to improve the forecasting accuracy
The set of alternatives is limited (e.g. to a few alternatives in scenario analysis)	All possible alternatives are sought

Product Development Phases

New product development (NPD) begins with the identification of market opportunity and ends with launching a product. Intermediate phases of the NPD process are variously distinguished in the literature. Cooper and Kleinschmidt (1986) describe in more detail the market and financial issues dividing the NPD process into initial screening, preliminary market assessment, preliminary technical assessment, detailed market study, business/financial analysis, product development, in-house product tests, customer tests of product, test market/trial sell, pre-commercialisation business analysis, production start-up, and market launch. Ulrich and Eppinger (2012) consider technical issues of product development with the following phases of the NPD process: planning, concept development, system-level design, detail design, tests and refinement, and production ramp-up. In turn, Crawford and Benedetto (2011) distinguish five phases in the NPD model: opportunity identification and selection, concept generation, concept/project evaluation, development (including both technical and marketing activities), and launch.

Each product development phase is preceded by a review regarding the completion of a previous phase, and the decision whether an NPD project should be or should not be continued. This study focuses on decision support in areas of selecting the most preferable concepts to a portfolio and resource allocation between NPD projects. Figure 1.6 illustrates product development phases from market analysis to product commercialisation with reference to decisions required between these phases. Depending on product properties, the product design phase requires iterations within detail design and prototype tests. As the presented study focuses on decisions regarding concept selection and resource allocation, the phases of concept evaluation and product design are further described in more detail.

Concept evaluation is concerned with assessing concepts according to evaluating criteria and selecting the best variant(s) of new product(s) taking into account a company's business strategy and available resources. This activity differs in companies depending on their procedures and the type of a new product. Companies more or less follow a sequence from quick looks to complete discounted cash flows and a new present value (Crawford and Benedetto 2011). A quick look approach is useful if a set of new product concepts is huge and a detailed consideration of each concept

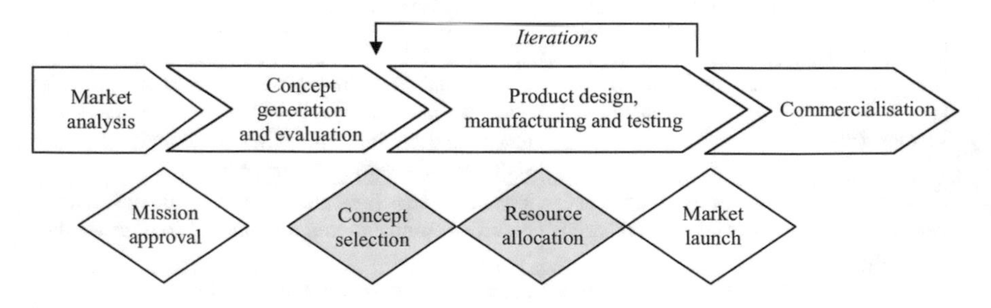

Fig. 1.6 Phases and related decisions in the NPD process

would be ineffective. Concept evaluation can use end-user screening and/or technical screening, depending on features of the concept. This process is more extensive in the case of a high degree of product newness. Acquiring information from potential customers about their perceptions related to the considered product concept can significantly improve the success of a new product. In turn, technical screening is concerned with determining the possibility of achieving the expected product utility, reliability or safety, and expenditures related to them. The process of concept evaluation ends with the decision about selecting the most promising concepts that can then be considered as NPD projects. Concept selection often uses a scoring model that can compare potential profits and costs in the expecting product life cycle.

The first phases of product development (market analysis, concept generation, and concept evaluation) compose the fuzzy front end of the NPD process. Most evaluations are based on imprecise and uncertain information about, for example, the time and cost of an NPD project, the unit production cost, the expected sales volume, and product life cycle. The required information depends on the degree of product newness and can be acquired from data referring to past experiences (e.g. the completed NPD projects) and/or new market research. The closer the NPD project completion is, the less uncertainty is in evaluations.

Concept selection is one of the most important decisions in the NPD process, taking into account the cost of developing an unsuccessful product. Moreover, the company's resources (e.g. the number of employees, financial means) are usually limited, imposing selection only on the most promising NPD projects. On the other hand, a new product with large potential may be omitted, and an enterprise can lose the potential profits.

Figure 1.7 presents a set of product concepts and the selection process in the case of one criterion, i.e. the NPD cost. The diameter of a circle corresponds to the estimated NPD cost. In turn, the width of the gate for the NPD cost depends on the R&D budget. Consequently, only concepts with an acceptable R&D cost will be considered for further development. However, the set of product concepts may include some concepts (grey circles in Fig. 1.7) that have the potential (e.g. from the strategic point of view) but that slightly exceed some project constraints (e.g. the NPD cost). The decision-maker may be interested in obtaining information about prerequisites that enable fulfilment of selection criteria (e.g. the cost of an NPD project, the profit from a new product) and/or the extent of changes of constraints to ensure the successful development of specific product concepts. To provide the above-mentioned information to the decision-maker, the problem specification in an inverse form has been proposed.

Product design is concerned with developing a physical product, including engineering design, manufacturing, and prototype tests. These activities are often not executed sequentially, but with overlapping to accelerate time to market for a new product and ensure its reliability, safety, and functionality. After preparing prototypes of a new product, the company can investigate customers' perceptions of product utility and quality.

Engineering design is based on product description (also called product protocol or product definition) that specifies product attributes desired by customers. Product

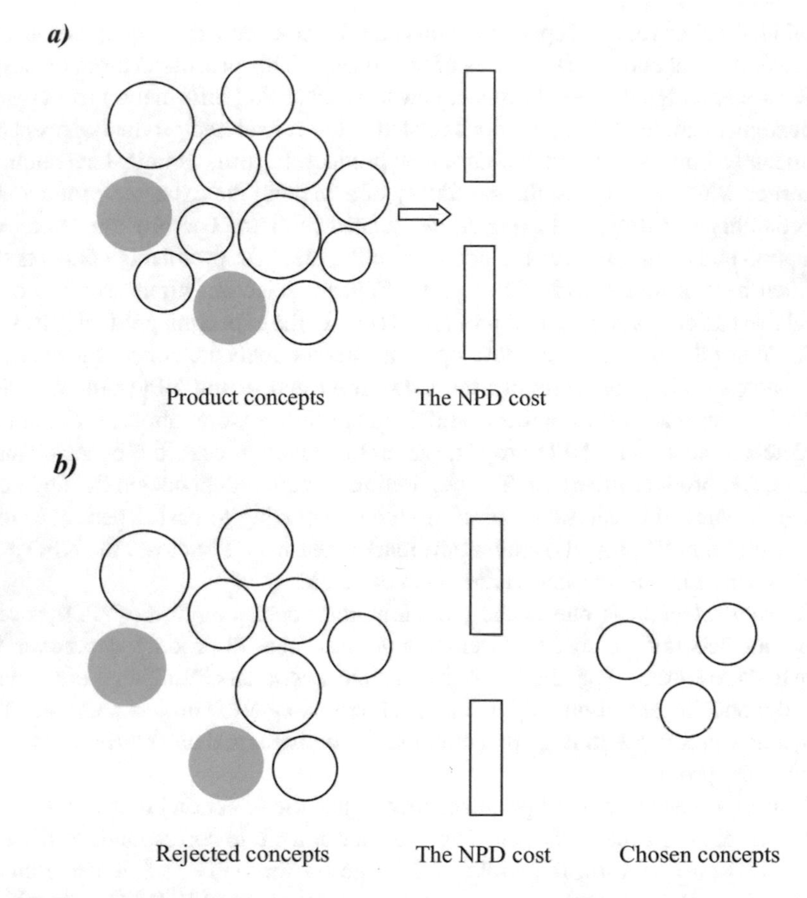

Fig. 1.7 Product concepts before selection (**a**) and after selection (**b**)

description is an agreement between the various stakeholders related to an NPD project, before extensive technical works will be initiated, and it can be seen as a trade-off between customers' needs and company's limitations (e.g. employees' qualification, available technology).

Product design is often preceded by resource preparation (e.g. materials, technologies, employees) depending on the degree of product newness. If a company is going to develop a new-to-the world product, then the team may need a suitable reward system, and the R&D department—additional financial outlays for testing numerous prototypes of a new product. A suitable resource preparation can accelerate product design and reduce resource wastage.

The process of product design includes uncertainties related to technical difficulties (e.g. in developing required technology and making prototypes) and customers' acceptance of product utility, quality, and price. Companies try to improve the chance of developing a successful product through periodic consumer research. This research aims to develop a new product that meets the desires of potential clients, prepare a preliminary marketing budget, or terminate an NPD project if market testing indicates that the product can be unprofitable. The phase of product design ends with business analysis that includes reliable cost estimates of manufacturing a new product and advertising campaign. The financial analysis should assure management that a new product is profitable enough to launch.

Resource allocation between NPD projects belongs to critical processes in product development. Resource allocation appears mainly in the product design phase, in which employees, machines, and financial means must be divided into NPD projects. If a new product cannot meet the requirements, then its development is stopped and available resources can be reallocated between ongoing projects. The proposed approach supports the decision-maker in identifying the possible alternatives of project completion, taking into account resource allocation.

Portfolio Management in Product Development

The number of potential NPD projects is usually greater than available resources in a company. As a result, NPD projects compete for scarce resources and require selection according to certain criteria (e.g. potential profitability). Project portfolio selection is the periodic activity that aims to select a set of NPD projects from project alternatives (both project proposals and projects currently underway), in which the company's objectives are met in a desirable manner without exceeding available resources and other constraints (Archer and Ghasemzadeh 1999).

A portfolio of new products is a set of NPD projects managed together to achieve a company's strategic objectives. The ultimate goal of project portfolio management is to maximise the contribution of NPD projects to business success (Heising 2012). Hence, project portfolio management involves the simultaneous management of the set of projects that constitute the company's investment strategy (Levine 2005; Patanakul and Milosevic 2009). Cooper et al. (1999) define portfolio management as a dynamic decision process, in which a business's list of active NPD projects is constantly revised and updated. This process is concerned with evaluating and prioritising new and existing projects to identify whether they should be developed or terminated, as well as allocating and reallocating resources to the active projects.

Meskendahl (2010) proposes a framework for achieving business success through specifying strategic orientation, project portfolio structuring, and project portfolio success. He describes project portfolio structuring with the use of consistency, integration, formalisation, and diligence. In turn, project portfolio success is considered in terms of average single project success, use of synergies, strategic fit, and portfolio balance. Successful NPD portfolios include a limited number of carefully selected, positioned, and balanced projects (Cooper et al. 2000). A balanced portfolio embraces a suitable distribution of projects taking into account technology and market risk, completion time, and return on investment. If a portfolio consists of too many NPD projects, there can appear conflict over existing resources, resulting in slowing project progress and reducing successful completion rates, e.g. by missed market opportunities (Archer and Ghasemzadeh 2007).

The importance of project portfolio management lies in evaluating, prioritising, and selecting projects according to the overall business strategy (Archer and Ghasemzadeh 2007; Meskendahl 2010). Many authors consider project prioritisation and resource allocation between ongoing projects as a key factor of success in project portfolio management (Cooper et al. 1999; Elonen and Artto 2003; Engwall and Jerbrant 2003; Blichfeldt and Eskerod 2008). Moreover, portfolio control is commonly seen as an important component of project portfolio management in coordinating and keeping the portfolio on track (Müller et al. 2008). Jonas (2010) expands the project portfolio management process towards organisational learning and portfolio exploitation that follows after portfolio structuring, resource management, and portfolio control.

The continuous evaluation of ongoing projects requires adjustment of measurement criteria to the project phase. For example, a project's risk profile related to project budget, completion time, resource demand, etc., is different at early and late phases of an NPD project. Consequently, the selection of a new project to the portfolio should take into account changes that have taken place after project initiation (Engwall and Jerbrant 2003). The initial and further evaluation of projects requires high-quality, up-to-date internal and external information that can make substantial effort in an organisation (Kaiser et al. 2015). This effort may be reduced through using computational techniques that are able to identify complex relationships from enterprise databases and use these relationships to a more precise estimation of project performance.

Figure 1.8 illustrates the actual cost for three ongoing NPD projects at the moment of evaluation t and cost estimation for ongoing projects and product concepts from the moment of time t. The aim of the proposed approach is to support the decision-maker in evaluating the potential of product concepts and selecting the most preferable portfolio of new products. Moreover, there is predicted future

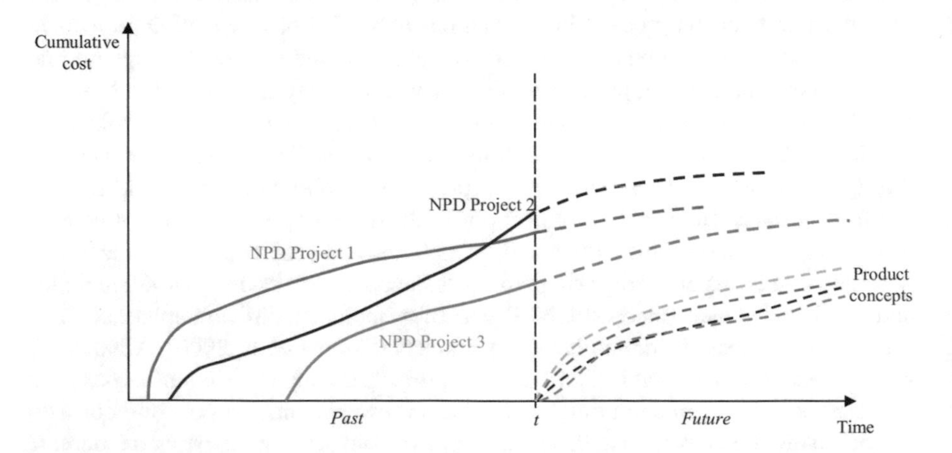

Fig. 1.8 Evaluation of ongoing NPD projects and product concepts

performance of ongoing products, and identified resource allocation such that it ensures optimal execution of NPD projects. Consequently, the proposed approach considers two problems: portfolio selection of product concepts and resource allocation in a multi-project environment.

The next subsections present a literature review of project selection criteria and methods for evaluating the potential of product concepts.

Project Selection Criteria

In general, evaluation criteria are not assigned to any particular project selection method. The evaluation of an NPD project using the same set of criteria aims to eliminate unfair competition between projects that may happen in the case of different reasoning for each comparison (Dutra et al. 2014). Table 1.2 presents a summary of project selection criteria in areas such as cost, technical difficulty, business benefits, and strategic benefits related to an NPD project.

The literature presents various studies addressing project selection and prioritisation (Henriksen and Traynor 1999; Cooper et al. 2001; Poh et al. 2001; Meade and Presley 2002; Verbano and Nosella 2010; Dutra et al. 2014; Parvaneh and El-Sayegh 2016). However, there is no consensus on what criteria should be used to make correct decisions regarding potential projects. As a result, each organisation tends to select a set of criteria that seem to be the most important. Nevertheless, this set can be incomplete or insufficient to support well-founded decisions. The wrong selection of decision criteria can lead the organisation to a failure in achieving its own and shareholders' strategic goals (Dutra et al. 2014). Consequently, there is a need to develop an approach that allows the decision-makers to obtain suitable decision criteria and their corresponding weights. These tasks can be successfully performed using computational intelligence techniques.

Project Selection Methods

Financial methods (return on investment—ROI, payback period—PP, net present value—NPV, internal rate of return—IRR, etc.) mainly employ financial criteria (e.g. capital budgeting techniques) to select NPD projects to a portfolio. The use of financial methods requires estimation of cash inflows and outflows, sometimes incorporating risk level related to a single project or entire project portfolio. The greatest limitations in using financial methods are reliable predictions of new product inflows and outflows. However, financial methods are the most popular project management and project selection approaches. According to Cooper et al. (2001), about three-fourth businesses use financial methods treating them as the dominant portfolio method. This can be a consequence of considering financial reasons (the

Table 1.2 Summary of project selection criteria appearing in the literature

Area	Selection criteria	References
Costs	Total investment	Badri et al. (2001), Lee and Kim (2001), Chien (2002), Linton et al. (2002), Meade and Presley (2002), Blau et al. (2004), Sun and Ma (2005), Medaglia et al. (2007), Eilat et al. (2008), Mavrotas et al. (2008), Kumar et al. (2009), Asosheh et al. (2010), Büyüközkan and Öztürkcan (2010), Gutjahr et al. (2010), Tohumcu and Karasakal (2010), Relich (2010b, 2016a, b), Alzahrani and Emsley (2013), Khalili-Damghani et al. (2013)
	Uncertainties involved	Henriksen and Traynor (1999), Badri et al. (2001), Meade and Presley (2002), Eilat et al. (2008), Kumar et al. (2009), Asosheh et al. (2010), Büyüközkan and Öztürkcan (2010), Chan and Ip (2010), Tohumcu and Karasakal (2010), Vidal et al. (2011), Khalili-Damghani et al. (2013), Huang and Zhao (2014), Relich (2016b), Relich and Pawlewski (2017)
Technical difficulty	Project complexity	Chien (2002), Blau et al. (2004), Coldrick et al. (2005), Eilat et al. (2008), Tohumcu and Karasakal (2010), Vidal et al. (2011), Alzahrani and Emsley (2013), Relich (2015, 2016b)
	Time pressure	Badri et al. (2001), Eilat et al. (2008), Asosheh et al. (2010), Tohumcu and Karasakal (2010), Kaiser et al. (2015), Relich (2010b)
	Degree of innovation	Hsu et al. (2003), Duarte and Reis (2006), Mavrotas et al. (2008), Yang and Hsieh (2009), Vidal et al. (2011), Khalili-Damghani et al. (2013), Kaiser et al. (2015), Relich (2015)
Business benefits	Market potential	Henriksen and Traynor (1999), Lee and Kim (2001), Linton et al. (2002), Meade and Presley (2002), Blau et al. (2004), Medaglia et al. (2007), Eilat et al. (2008), Kumar et al. (2009), Yang and Hsieh (2009), Asosheh et al. (2010), Büyüközkan and Öztürkcan (2010), Chan and Ip (2010), Tohumcu and Karasakal (2010), Kaiser et al. (2015), Relich (2015), Relich and Pawlewski (2017)
	Overall benefits	Coldrick et al. (2005), Kumar et al. (2009), Yang and Hsieh (2009), Büyüközkan and Öztürkcan (2010), Vidal et al. (2011), Relich (2015), Morton et al. (2016), Relich and Pawlewski (2017)
	Meeting customers' needs	Badri et al. (2001), Eilat et al. (2008), Mavrotas et al. (2008), Asosheh et al. (2010), Chan and Ip (2010), Tohumcu and Karasakal (2010), Kaiser et al. (2015), Relich (2016b), Relich and Pawlewski (2017)

(continued)

Table 1.2 (continued)

Area	Selection criteria	References
Strategic benefits	Competitiveness improvement	Henriksen and Traynor (1999), Duarte and Reis (2006), Eilat et al. (2008), Mavrotas et al. (2008), Tohumcu and Karasakal (2010), Kaiser et al. (2015), Relich (2016b)
	Strategic alignment	Henriksen and Traynor (1999), Meade and Presley (2002), Eilat et al. (2008), Asosheh et al. (2010), Vidal et al. (2011), Khalili-Damghani et al. (2013), Relich and Pawlewski (2017)
	Intangible benefits	Badri et al. (2001), Duarte and Reis (2006), Asosheh et al. (2010), Gutjahr et al. (2010), Vidal et al. (2011), Alzahrani and Emsley (2013), Relich (2015)

maximisation of returns, R&D productivity) as the most important factor in portfolio management.

Mathematical methods (linear programming, non-linear programming, integer programming, goal programming, data envelopment analysis—DEA) optimise a specific objective function (e.g. the benefit expected from an NPD project portfolio by resource constraints). Unlike the simple linear and non-linear programming models, goal programming enables weight assignment for goals specified by the decision-maker. Goal programming uses a set of yes/no choices to obtain information whether an NPD project should be initiated, continued, or terminated. DEA is a linear programming methodology that calculates the relative efficiency of multiple decision-making units on the basis of observed inputs and outputs, which may be expressed with different types of metrics (Eilat et al. 2008). These methods require an extensive range of highly complex input data, and they are based on sophisticated algorithms that derive from operational research (Verbano and Nosella 2010).

NPD project selection can also be supported through *computational intelligence*, including artificial neural networks, genetic algorithms, fuzzy logic, or hybrid systems such as neuro-fuzzy systems. Artificial neural networks are successfully applied to estimation of monetary criteria for evaluating an NPD project, e.g. cost or sales revenue related to a new product (Thiesing and Vornberger 1997; Boussabaine and Kaka 1998). Genetic algorithms can facilitate searching the optimal sequence of product dependencies and limited resources (Blau et al. 2004) or generating alternatives within a multi-objective framework (Khalili-Damghani et al. 2013). In turn, fuzzy logic is often used for specification of non-monetary criteria of project evaluation. The relative importance of these criteria is mainly evaluated by experts. The need for retaining knowledge from experts in an organisation results in the development of expert systems that consist of knowledge base, inference engine, and user interface, in which non-expert users can formulate questions and receive answers.

Decision analysis (e.g. Analytic Hierarchy Process—AHP, Multi-Attribute Utility Technique—MAUT, decision trees) is a decision-making model that selects the best projects by designing a tiered framework, in which project alternatives are placed on the bottom level and objectives are on higher levels. The alternatives are compared and the best projects are selected through constructing matrices that summarise the priorities on each level. To involve monetary and non-monetary factors in evaluating projects, fuzzy AHP is often used (Büyüközkan and Feyzioğlu 2004b; Enea and Piazza 2004). MAUT calculates a utility function for each project, assigning individual scores to attributes and evaluating overall project utility. The main characteristic of this method is its ability to break down a complex problem into a set of sub-problems that can be solved individually in accordance with multiple objectives. More sophisticated techniques are used to deal with interdependent projects. In turn, decision trees are used if a project can be specified by a sequence of decisions, where each decision depends on the outcome of the previous decision (Verbano and Nosella 2010). Decision trees are often combined with other methods in order to improve their applicability.

Scoring methods specify a set of criteria that are used for project selection. Projects are rated on a number of criteria, for example: low-medium-high, 1–5 or 0–10 scales. Each project obtains a score that expresses the extent to which a set of criteria were met. Checklists are the simplest method, in which projects are evaluated on a set of yes/no questions. Each project should achieve either all yes answers or a certain number of yes answers to proceed (Cooper et al. 2001). Scoring algorithms use an additive or multiplicative algorithm to summarise the opinions expressed within peer reviews, which are compiled by filling out a questionnaire. An algorithm enables weight assignment for individual criterion in order to emphasise its importance. The overall objective function is calculated as the sum of the value of each project. These methods use subjective input data and procedures that vary in sophistication to provide numerical output data (Verbano and Nosella 2010).

Interactive methods (e.g. Delphi) compare each project solely on the basis of subjective evaluations without using mathematical algorithms. The decision-making process is conducted by comparing the opinions of the involved actors about each project. The most popular method within interactive methods is Delphi that involves a panel of experts, who answer a series of structured questionnaires and prepare feedback reports on the subject. These methods are often carried out repeatedly until a satisfactory level of agreement is reached.

In *bubble diagrams*, projects are located as circles on an X–Y plot that includes dimensions such as probability of technical success and reward (NPV). Circle size refers to the annual amount of resources. As a result, projects are categorised according to the zone or quadrant they are in (Cooper et al. 2001). The *balanced scorecard* facilitates a shift from financial based evaluation techniques to strategy and vision (Milis and Mercken 2004). The balanced scorecard consists of a collection of measures arranged in groups (cards) that offer an evaluation of the organisational performance along financial, marketing, operational, and strategic dimensions (Eilat et al. 2008).

The above methods differ in structural characteristics regarding the type of input data, the data analysis process, and the type of output data that can be obtained. Consequently, they belong with different degree to quantitative and qualitative approaches. Methods belonging to a quantitative approach use quantitative input data and adopt strict procedures (e.g. mathematic algorithms) for obtaining a quantitative output on the basis of financial and economic indices. In turn, methods belonging to a qualitative approach use only qualitative data referring to the comparison of opinions of key actors involved in the NPD process to obtain a qualitative output for selecting projects. There are many intermediate forms between quantitative and qualitative approaches. For example, a semi-quantitative approach differs from a quantitative approach in the context of subjective evaluation that is used in the selection procedure, whereas in a semi-qualitative approach qualitative data are used and the data analysis process towards obtaining a quantitative output is adopted. Table 1.3 presents a summary of project selection methods from a quantitative to qualitative approach.

The above-presented methods allow the decision-makers to obtain information about the potential of a product concept according to evaluation criteria. These methods enable us to answer a question stated in *forward* form, e.g. 'what profit is related to an NPD project for given values of assumed criteria?' The proposed approach is concerned with the provision of information about prerequisites that must be met to ensure a preferable profit of a new product. This is related to a question stated in *inverse* form (see Fig. 1.4).

Methods for Evaluating the Potential of a New Product

The potential of a new product can be measured as the comparison of predicted cash inflows with outflows within a given period of time. Consequently, there is a need to predict sales and costs of product development, manufacturing, and marketing. Sales forecasting and cost estimation depend on available data regarding similar completed NPD projects. If a new product belongs to the existing product line, then analogical models can be used for cost estimation. In turn, if a company develops an entirely new product, then analytical models are more useful for evaluating sales or costs of the new product. Figure 1.9 illustrates cumulative cash flows (sales revenue, costs, and corresponding profit/loss).

Cumulative costs include the cost of product development, marketing, and production. Time to market refers to product development, from market analysis to launch. The comparison of alternative new products requires the prediction of product lifespan and potential sales and costs in this period. The evaluation of the potential of new products can also include the discount rate (as in the NPV method) towards reflecting opportunity cost of capital and inflation in the investment period.

The cost estimation of marketing and production is based on more certain factors than product development. For instance, the marketing budget can be equal for each new product, and production costs depend on factors such as materials, labour, and sales volumes. Consequently, the evaluation of product attractiveness is related to cost estimation of product development, sales forecasting, and lifespan prediction for a new product.

Ben-Arieh and Qian (2003) divide cost estimation methods into four groups: intuitive, analogical, parametric, and analytical. Intuitive methods use past experience of an estimator. Analogical methods estimate the cost of new products using similarity to previous similar products. Parametric methods estimate the cost of a

Table 1.3 Summary of project selection methods appearing in the literature

Method	References
Financial indices (ROI, PP, NPV, IRR, etc.)	Cooper et al. (2001), Poh et al. (2001), Blau et al. (2004), De Reyck et al. (2008), Wiesemann et al. (2010), Dutra et al. (2014), Guerra et al. (2014), Leyman and Vanhoucke (2016)
Linear programming	Heidenberger and Stummer (1999), Chien (2002), Gutjahr et al. (2010), Parvaneh and El-Sayegh (2016)
Non-linear programming	Loch and Kavadias (2002), Blau et al. (2004), Medaglia et al. (2007), Carazo et al. (2010), Gutjahr et al. (2010)
Integer programming	Archer and Ghasemzadeh (1999), Shirland et al. (2003), Sun and Ma (2005), Tavana et al. (2015)
Goal programming	Badri et al. (2001), Lee and Kim (2001), Shirland et al. (2003)
Artificial neural networks	Thieme et al. (2000), Yazgan et al. (2009), Relich (2010b), Costantino et al. (2015), Relich and Pawlewski (2017)
Genetic algorithms	Blau et al. (2004), Medaglia et al. (2007), Bhattacharyya et al. (2011), Khalili-Damghani et al. (2013)
Neuro-fuzzy systems	Lin and Yeh (2001), Büyüközkan and Feyzioğlu (2004a), Jin et al. (2007), Relich (2010a, 2016b)
Fuzzy logic	Wang and Hwang (2007), Bhattacharyya et al. (2011), Chang and Lee (2012), Ghapanchi et al. (2012), Lin and Yang (2015), Relich (2015), Yan and Ma (2015), Relich and Pawlewski (2017), Pérez et al. (2018)
Expert systems	Liberatore (1988), Chu et al. (1996), Archer and Ghasemzadeh (1999), Tian et al. (2002), Relich et al. (2015), Relich and Pawlewski (2015), Relich (2016a)
Data envelopment analysis (DEA)	Linton et al. (2002), Eilat et al. (2008), Asosheh et al. (2010), Tohumcu and Karasakal (2010), Chang and Lee (2012), Ghapanchi et al. (2012), Khalili-Damghani et al. (2013), Karasakal and Aker (2017)
Analytic network process (ANP)	Lee and Kim (2001), Meade and Presley (2002), Cheng and Li (2005), Mohanty et al. (2005), Yazgan et al. (2009), Büyüközkan and Öztürkcan (2010), Tohumcu and Karasakal (2010)

(continued)

Table 1.3 (continued)

Method	References
Analytic hierarchy process (AHP)	Poh et al. (2001), Al Khalil (2002), Cheng et al. (2002), Enea and Piazza (2004), Cebeci (2009), Parvaneh and El-Sayegh (2016), Karasakal and Aker (2017)
Decision trees	Poh et al. (2001), De Reyck et al. (2008)
Multi-attribute utility technique (MAUT)	Bose et al. (1997), Duarte and Reis (2006), Wallenius et al. (2008), Lopes and de Almeida (2015)
Scoring models	Henriksen and Traynor (1999), Poh et al. (2001), Coldrick et al. (2005), Mavrotas et al. (2008), Kumar et al. (2009), Relich and Pawlewski (2017)
Delphi	Lee and Kim (2001), Yang and Hsieh (2009)
Bubble diagrams	Cooper et al. (2001), Blau et al. (2004)
Balanced scorecard (BSC)	Milis and Mercken (2004), Eilat et al. (2008), Cebeci (2009), Asosheh et al. (2010), Chan and Ip (2010)

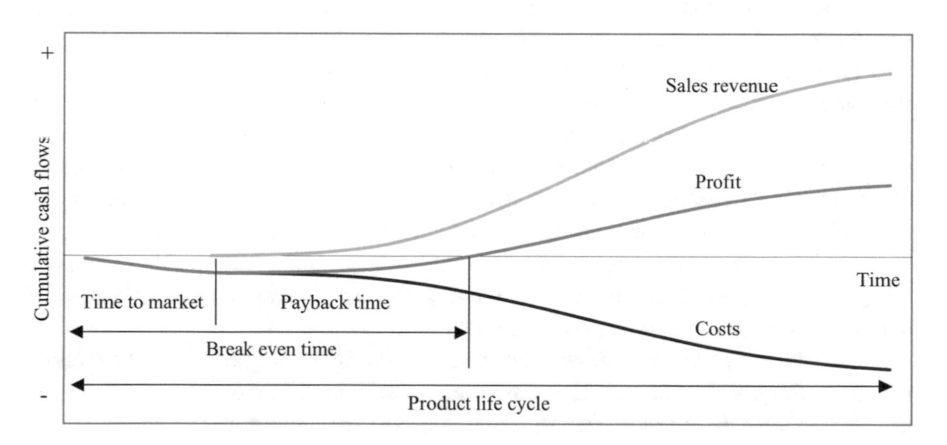

Fig. 1.9 Product development cash flow

new product from parameters that significantly influence the cost. In turn, analytical methods estimate the cost of a product using a decomposition of the work into elementary tasks with known cost.

Cavalieri et al. (2004) categorise cost estimation techniques into analogy-based techniques, parametric models, and engineering approaches. Analogy-based techniques belong to qualitative estimation methods, and they are based on similarity analysis between a new product and existing products. Parametric models include an analytical function of a set of variables that are usually related to some features of a new product (e.g. dimensions, used materials) and that are supposed to have a significant impact on NPD project performance. Engineering approaches base on the detailed analysis of product features and manufacturing process. For example, cost estimation of a new product is calculated in this approach as the sum of resources used to design and produce each product component (e.g. raw materials, labour, equipment). As a result, the engineering approach is suitable in the final product development phases, in which the product and manufacturing process are well defined.

Niazi et al. (2006) classify product cost estimation techniques into qualitative and quantitative. Qualitative techniques are mainly based on a comparison analysis of a new product with previous products in order to identify similarities between them. If design and manufacturing processes are similar both in past products and in a new product, the identified relationships can reduce the cost of obtaining estimates. In turn, quantitative techniques refer to a detailed analysis of a product design, its features, and corresponding manufacturing processes instead of relying on the past data or knowledge of an estimator. Thus, cost estimation is based on an analytical function of certain variables representing different product parameters or the sum of elementary units representing different resources consumed during product development.

According to Niazi et al. (2006), qualitative techniques include regression analysis, back-propagation neural networks, case-based reasoning, rule-based system, fuzzy logic system, and expert system. In turn, quantitative techniques are divided into parametric cost estimation and techniques referring to analytical cost estimation such as operation-based approach, breakdown approach, tolerance-based cost models, feature-based cost estimation, and activity-based cost estimation. Analytical cost estimation techniques require decomposing a product into elementary units and manufacturing operations. For example, an operation-based approach enables the estimation of manufacturing cost taking into account material cost, factory expenses, set-up time, operation time, and non-operation time (Jung 2002). Consequently, analytic techniques are mainly used in the final design phases, in which information about a new product is more precisely specified.

The NPD project selection problem refers to the beginning of product development. Consequently, analogical and parametric estimation methods are more suitable to evaluate the NPD project duration and cost than analytical techniques. The proposed methodology is based on the use of some computational intelligence techniques (artificial neural networks and neuro-fuzzy structures) in the context of parametric models. Moreover, cause-and-effect relationships identified by these techniques can be used to solve the problem stated in the inverse form through generating scenarios regarding possible solutions to the NPD-related problem.

Scenario Analysis in the NPD Process

Scenarios are seen as different possible future states of a system (Tietje 2005). Scenario analysis has become a major decision tool in economics and strategic management (Borgonovo 2017). A wide range of literature refers to the methodological aspects of scenario generation. O'Brien presents (2004) the scenario development process as a top-down, qualitative sequence of the following steps: (1) set the scene, (2) generate uncertain and predetermined factors, (3) reduce factors and specify factors ranges, (4) choose themes and develop scenario details, (5) check consistency of scenarios, (6) present scenarios, (7) assess impact of scenarios, and (8) develop and test strategies. From a mathematical point of view, scenario analyses can be divided into holistic scenarios, model scenarios, and formative scenarios (Tietje 2005). A holistic scenario analysis develops scenarios by experts in the required disciplines. A model scenario analysis explicitly uses a (not always dynamic) system model such as an economic or environmental model (Costanza et al. 1993). In turn, a formative scenario analysis is based on qualitatively assessed impact factors and expert-rated quantitative relationships between them. In this classification, 'formative' indicates the generic mathematical structure behind the scenarios that is linked with quantitative/qualitative expert assessments (Tietje 2005).

Tietje (2005) argues that a scenario within consistency analysis can be conceived as a set of system variables (impact factors). A considerable number of impact variables may cause the large set of scenarios. As a result, there is a need to determine a trade-off between the reduction of the number of scenarios and the quality of scenario analysis. This quality may be considered from the perspective of consistent scenarios, diverse scenarios, a reliable set of scenarios, and efficient scenarios (i.e. the most consistent scenarios within a group of similar scenarios, because they are the most relevant representatives). Moreover, Heugens and van Oosterhout (2001) indicate that the small number of scenarios is advantageous for decision-makers who can hardly compare many qualitatively different scenarios.

Scenario analysis is widely used by practitioners who are interested in comparing different scenarios that can be seen as possible states of a system. Since these states result from the values of model inputs, creating a reliable model of a project-enterprise-environment system is the task of primary importance. In terms of quantitative modelling, scenario analysis supports decision-makers in evaluating the variability of predictions (Borgonovo 2017).

Scenario analysis in the context of NPD aims to evaluate the potential variability in the NPV related to an NPD project. This analysis computes several net present values for the NPD project based on different scenarios. The scenario with the most expected cash flows is called the basic scenario (variant) that is compared to other scenarios. Typically there are developed at least two scenarios for a pessimistic and optimistic case that allow decision-makers to perceive a likely range of the project's NPV. Moreover, scenario analysis examines the joint impact on the NPV of simultaneous changes in several elements (variables) occurring in an NPD model.

Compared to the basic scenario, the pessimistic scenario includes lower sales volume, lower price of a new product, shorter product life cycle, higher costs, etc. The pessimistic scenario reflects the circumstances related to entire disaster of an NPD project, and can use the data referring to past project failures. In turn, the optimistic scenario should illustrate product development if everything is better than expected. However, optimistic boundaries of sales volume, price, costs, and so on should base on reasonable optimism. An unrealistic scenario contributes little to the analysis, and it may even lead to harmful decisions (Lee and Lee 2006).

Scenario analysis simultaneously modifies many variables that impact cash flows and the NPV to build different scenarios. In turn, sensitivity analysis changes one variable at a time from its basic value. This presents changes in the NPV according to changes in individual variables (Lee and Lee 2006). In sensitivity analysis, the values of input variables of an NPD model are changed by, for example, 10%, to assess changes in the NPV. Decision-makers may be particularly interested in identifying variables that significantly impact the NPV, especially if the NPV of an NPD project is negative for some reasonable values of these variables. The results of sensitivity analysis may be illustrated in a tornado diagram that provides an intuitive and easy-to-interpret graphical representation of sensitivities related to factors occurring in the NPV analysis (Borgonovo 2017).

Scenario analysis provides different variants of the NPV using the analyst's estimates of expected cash flows for specific changes in model parameters (Lee and Lee 2006). Moreover, current scenario analysis is carried out within a forward approach, i.e. the predicted values of variables in an NPD model are used to compute the NPV. The proposed approach expands current research in scenario analysis towards using computational intelligence techniques to identify cause-and-effect relationships among enterprise databases and apply these relationships to express the joint impact of all variables on the NPV (the change of one variable causes changes in others). Moreover, the proposed approach enables problem specification in the inverse form allowing the decision-maker to verify quickly whether the desirable NPV of an NPD project exists or not; if there is a set of solutions (variants that achieve the desirable NPV), then the decision-maker can obtain all admissible combinations of chosen variables of an NPD model. Furthermore, the proposed approach provides foundations for using a parametric model to estimate sales volume and costs of product development, production, and marketing. Consequently, parameters occurring in a parametric model (related to product and enterprise) can be used to manage NPD project performance towards achieving the desirable NPV. The current scenario analysis provides information about desirable changes in costs and sales volume, but it seems to be too general and abstract to manage the NPD project towards a desirable outcome. The comparison of traditional scenario analysis with the proposed approach in the context of their advantages and limitations is presented in Chap. 5.

Summary

Product development is a complex process that involves various factors placed in an enterprise and its environment. Moreover, product development is a crucial issue for maintaining competitiveness, reducing business profits in the case of promoting unsuccessful products. As a result, the decision-maker should be supported towards obtaining possible variants of NPD project performance and their feasibility within specified constraints. This study is concerned with NPD-related problems, including issues of project portfolio selection, evaluation of the potential of a new product, and resource reallocation between new and ongoing projects.

The literature review presents widely used methods for project selection and evaluation of the potential of a new product. Current methods have been developed for solving the project selection problem in the forward form, in which the optimal portfolio includes NPD projects with the greatest potential within specific selection criteria and constraints. However, there is a lack of methods for identifying prerequisites that enable the improvement of the attractiveness of NPD projects. The proposed approach fills this gap, allowing the decision-maker to recognise alternatives of product development within the specified constraints. The problem formulated in the inverse form makes possible to search variants of NPD project performance that should meet desirable values of project selection criteria.

The proposed approach is dedicated to enterprises that develop new products through modifying previous products. Enterprise databases can then be used to the identification of relationships between input and output variables towards improving the precision of evaluating the potential of a new product. These relationships together with facts and constraints constitute a knowledge base that is the main part of a decision support system.

A systems approach embraces factors related to a product, enterprise, and its environment, enabling design of a logical process of solving problem. The specification of these factors in terms of variables and constraints facilitates statement and solution of NPD-related problems. As a result, a systems approach is a pertinent framework for generating variants of problem-solving. Advantageous features of a systems approach and problems stated in the inverse form lead towards a need of developing an NPD model in terms of a declarative representation.

References

Al Khalil, M. I. (2002). Selecting the appropriate project delivery method using AHP. *International Journal of Project Management, 20*(6), 469–474.

Alzahrani, J. I., & Emsley, M. W. (2013). The impact of contractors' attributes on construction project success: A post construction evaluation. *International Journal of Project Management, 31*(2), 313–322.

Archer, N. P., & Ghasemzadeh, F. (1999). An integrated framework for project portfolio selection. *International Journal of Project Management, 17*(4), 207–216.

Archer, N., & Ghasemzadeh, F. (2007). Project portfolio selection and management. In P. Morris & J. Pinto (Eds.), *The Wiley guide to project, program & portfolio management* (pp. 94–112). Hoboken, NJ: Wiley.

Asosheh, A., Nalchigar, S., & Jamporazmey, M. (2010). Information technology project evaluation: An integrated data envelopment analysis and balanced scorecard approach. *Expert Systems with Applications, 37*, 5931–5938.

Badri, M. A., Davis, D., & Davis, D. (2001). A comprehensive 0–1 goal programming model for project selection. *International Journal of Project Management, 19*(4), 243–252.

Ben-Arieh, D., & Qian, L. (2003). Activity-based cost management for design and development stage. *International Journal of Production Economics, 83*(2), 169–183.

Bhattacharyya, R., Kumar, P., & Kar, S. (2011). Fuzzy R&D portfolio selection of interdependent projects. *Computers & Mathematics with Applications, 62*(10), 3857–3870.

Blau, G. E., Pekny, J. F., Varma, V. A., & Bunch, P. R. (2004). Managing a portfolio of interdependent new product candidates in the pharmaceutical industry. *Journal of Product Innovation Management, 21*(4), 227–245.

Blichfeldt, B. S., & Eskerod, P. (2008). Project portfolio management–There's more to it than what management enacts. *International Journal of Project Management, 26*(4), 357–365.

Borgonovo, E. (2017). *Sensitivity analysis. International series in operations research and management science*. Cham: Springer.

Bose, U., Davey, A. M., & Olson, D. L. (1997). Multi-attribute utility methods in group decision-making: past applications and potential for inclusion in GDSS. *Omega, 25*(6), 691–706.

Boussabaine, A. H., & Kaka, A. P. (1998). A neural networks approach for cost flow forecasting. *Construction Management & Economics, 16*(4), 471–479.

Büyüközkan, G., & Feyzioğlu, O. (2004a). A fuzzy-logic-based decision-making approach for new product development. *International Journal of Production Economics, 90*(1), 27–45.

Büyüközkan, G., & Feyzioğlu, O. (2004b). A new approach based on soft computing to accelerate the selection of new product ideas. *Computers in Industry, 54*(2), 151–167.

Büyüközkan, G., & Öztürkcan, D. (2010). An integrated analytic approach for Six Sigma project selection. *Expert Systems with Applications, 37*(8), 5835–5847.

Carazo, A. F., Gómez, T., Molina, J., Hernández-Díaz, A. G., Guerrero, F. M., & Caballero, R. (2010). Solving a comprehensive model for multiobjective project portfolio selection. *Computers & Operations Research, 37*(4), 630–639.

Cavalieri, S., Maccarrone, P., & Pinto, R. (2004). Parametric vs. neural network models for the estimation of production costs: A case study in the automotive industry. *International Journal of Production Economics, 91*(2), 165–177.

Cebeci, U. (2009). Fuzzy AHP-based decision support system for selecting ERP systems in textile industry by using balanced scorecard. *Expert Systems with Applications, 36*(5), 8900–8909.

Chan, S. L., & Ip, W. H. (2010). A Scorecard-Markov model for new product screening decisions. *Industrial Management & Data Systems, 110*(7), 971–992.

Chang, P. T., & Lee, J. H. (2012). A fuzzy DEA and knapsack formulation integrated model for project selection. *Computers & Operations Research, 39*(1), 112–125.

Cheng, E., & Li, H. (2005). Analytic network process applied to project selection. *Journal of Construction Engineering and Management, 131*, 459–466.

Cheng, E., Li, H., & Ho, D. (2002). Analytic hierarchy process (AHP): A defective tool when used improperly. *Measuring Business Excellence, 6*(4), 33–37.

Chien, C. F. (2002). A portfolio–evaluation framework for selecting R&D projects. *R&D Management, 32*(4), 359–368.

Chu, P., Hsu, Y., & Fehling, M. (1996). A decision support system for project portfolio selection. *Computers in Industry, 32*(2), 141–149.

Coldrick, S., Longhurst, P., Ivey, P., & Hannis, J. (2005). An R&D options selection model for investment decisions. *Technovation, 25*(3), 185–193.

Cooper, R. G., & Kleinschmidt, E. J. (1986). An investigation into the new product process: Steps, deficiencies, and impact. *Journal of Product Innovation Management, 3*(2), 71–85.

Cooper, R. G., Edgett, S. J., & Kleinschmidt, E. J. (1999). New product portfolio management: Practices and performance. *Journal of Product Innovation Management, 16*(4), 333–351.

Cooper, R. G., Edgett, S. J., & Kleinschmidt, E. J. (2000). New problems, new solutions: Making portfolio management more effective. *Research-Technology Management, 43*(2), 18–33.

Cooper, R., Edgett, S., & Kleinschmidt, E. (2001). Portfolio management for new product development: Results of an industry practices study. *R&D Management, 31*(4), 361–380.

Costantino, F., Di Gravio, G., & Nonino, F. (2015). Project selection in project portfolio management: An artificial neural network model based on critical success factors. *International Journal of Project Management, 33*(8), 1744–1754.

Costanza, R., Wainger, L., Folke, C., & Mäler, K. G. (1993). Modeling complex ecological economic systems: Toward an evolutionary, dynamic understanding of people and nature. In *Ecosystem management* (pp. 148–163). New York: Springer.

Crawford, M., & Benedetto, A. D. (2011). *New products management* (10th ed.). New York: McGraw-Hill Education.

De Reyck, B., Degraeve, Z., & Vandenborre, R. (2008). Project options valuation with net present value and decision tree analysis. *European Journal of Operational Research, 184*(1), 341–355.

Duarte, B. P., & Reis, A. (2006). Developing a projects evaluation system based on multiple attribute value theory. *Computers & Operations Research, 33*(5), 1488–1504.

Dutra, C. C., Ribeiro, J. L., & de Carvalho, M. M. (2014). An economic–probabilistic model for project selection and prioritization. *International Journal of Project Management, 32*(6), 1042–1055.

Eilat, H., Golany, B., & Shtub, A. (2008). R&D project evaluation: An integrated DEA and balanced scorecard approach. *Omega, 36*, 895–912.

Elonen, S., & Artto, K. A. (2003). Problems in managing internal development projects in multi-project environments. *International Journal of Project Management, 21*(6), 395–402.

Enea, M., & Piazza, T. (2004). Project selection by constrained fuzzy AHP. *Fuzzy Optimization and Decision-Making, 3*(1), 39–62.

Engwall, M., & Jerbrant, A. (2003). The resource allocation syndrome: the prime challenge of multi-project management? *International Journal of Project Management, 21*(6), 403–409.

Ghapanchi, A. H., Tavana, M., Khakbaz, M. H., & Low, G. (2012). A methodology for selecting portfolios of projects with interactions and under uncertainty. *International Journal of Project Management, 30*(7), 791–803.

Guerra, M. L., Magni, C. A., & Stefanini, L. (2014). Interval and fuzzy average internal rate of return for investment appraisal. *Fuzzy Sets and Systems, 257*, 217–241.

Gutjahr, W. J., Katzensteiner, S., Reiter, P., Stummer, C., & Denk, M. (2010). Multi-objective decision analysis for competence-oriented project portfolio selection. *European Journal of Operational Research, 205*(3), 670–679.

Heidenberger, K., & Stummer, C. (1999). Research and development project selection and resource allocation: a review of quantitative modelling approaches. *International Journal of Management Reviews, 1*(2), 197–224.

Heising, W. (2012). The integration of ideation and project portfolio management – A key factor for sustainable success. *International Journal of Project Management, 30*(5), 582–595.

Henriksen, A. D., & Traynor, A. J. (1999). A practical R&D project-selection scoring tool. *IEEE Transactions on Engineering Management, 46*(2), 158–170.

Heugens, P. P., & van Oosterhout, J. (2001). To boldly go where no man has gone before: integrating cognitive and physical features in scenario studies. *Futures, 33*(10), 861–872.

Hsu, Y. G., Tzeng, G. H., & Shyu, J. Z. (2003). Fuzzy multiple criteria selection of government-sponsored frontier technology R&D projects. *R&D Management, 33*(5), 539–551.

Huang, X., & Zhao, T. (2014). Project selection and scheduling with uncertain net income and investment cost. *Applied Mathematics and Computation, 247*, 61–71.

Jin, H., Zhao, J., & Chen, X. (2007). The application of neuro-fuzzy decision tree in optimal selection of technological innovation projects. *Software Engineering, Artificial Intelligence, Networking, and Parallel/Distributed Computing, 3*, 438–443.

Jonas, D. (2010). Empowering project portfolio managers: How management involvement impacts project portfolio management performance. *International Journal of Project Management, 28*(8), 818–831.

Jung, J. Y. (2002). Manufacturing cost estimation for machined parts based on manufacturing features. *Journal of Intelligent Manufacturing, 13*(4), 227–238.

Kaiser, M. G., El Arbi, F., & Ahlemann, F. (2015). Successful project portfolio management beyond project selection techniques: Understanding the role of structural alignment. *International Journal of Project Management, 33*(1), 126–139.

Karasakal, E., & Aker, P. (2017). A multicriteria sorting approach based on data envelopment analysis for R&D project selection problem. *Omega, 73*, 79–92.

Kerzner, H. (2001). *Project management: A systems approach to planning, scheduling, and controlling* (7th ed.). New York: Wiley.

Khalili-Damghani, K., Sadi-Nezhad, S., Lotfi, F. H., & Tavana, M. (2013). A hybrid fuzzy rule-based multi-criteria framework for sustainable project portfolio selection. *Information Sciences, 220*, 442–462.

Kumar, M., Antony, J., & Rae Cho, B. (2009). Project selection and its impact on the successful deployment of six sigma. *Business Process Management Journal, 15*(5), 669–686.

Lee, J. W., & Kim, S. H. (2001). An integrated approach for interdependent information system project selection. *International Journal of Project Management, 19*(2), 111–118.

Lee, C. F., & Lee, A. C. (Eds.). (2006). *Encyclopedia of finance*. New York: Springer.

Levine, H. A. (2005). *Project portfolio management: A practical guide to selecting projects, managing portfolios, and maximizing benefits*. New York: Wiley.

Leyman, P., & Vanhoucke, M. (2016). Payment models and net present value optimization for resource-constrained project scheduling. *Computers & Industrial Engineering, 91*, 139–153.

Liberatore, M. J. (1988). An expert support system for R&D project selection. *Mathematical and Computer Modelling, 11*, 260–265.

Lin, C. T., & Yang, Y. S. (2015). A linguistic approach to measuring the attractiveness of new products in portfolio selection. *Group Decision and Negotiation, 24*(1), 145–169.

Lin, Y. C., & Yeh, J. M. (2001). A fuzzy controlled neural network for screening new product ideas. *Journal of Information and Optimization Sciences, 22*(1), 91–111.

Linton, J. D., Walsh, S. T., & Morabito, J. (2002). Analysis, ranking and selection of R&D projects in a portfolio. *R&D Management, 32*(2), 139–148.

Loch, C. H., & Kavadias, S. (2002). Dynamic portfolio selection of NPD programs using marginal returns. *Management Science, 48*(10), 1227–1241.

Lopes, Y. G., & de Almeida, A. T. (2015). Assessment of synergies for selecting a project portfolio in the petroleum industry based on a multi-attribute utility function. *Journal of Petroleum Science and Engineering, 126*, 131–140.

Mavrotas, G., Diakoulaki, D., & Kourentzis, A. (2008). Selection among ranked projects under segmentation, policy and logical constraints. *European Journal of Operational Research, 187*(1), 177–192.

Meade, L. M., & Presley, A. (2002). R&D project selection using the analytic network process. *IEEE Transactions on Engineering Management, 49*(1), 59–66.

Medaglia, A. L., Graves, S. B., & Ringuest, J. L. (2007). A multiobjective evolutionary approach for linearly constrained project selection under uncertainty. *European Journal of Operational Research, 179*(3), 869–894.

Meskendahl, S. (2010). The influence of business strategy on project portfolio management and its success – A conceptual framework. *International Journal of Project Management, 28*(8), 807–817.

Milis, K., & Mercken, R. (2004). The use of the balanced scorecard for the evaluation of information and communication technology projects. *International Journal of Project Management, 22*(2), 87–97.

Mohanty, R. P., Agarwal, R., Choudhury, A. K., & Tiwari, M. K. (2005). A fuzzy ANP-based approach to R&D project selection: a case study. *International Journal of Production Research, 43*(24), 5199–5216.

Morton, A., Keisler, J. M., & Salo, A. (2016). Multicriteria portfolio decision analysis for project selection. In S. Greco, M. Ehrgott, & J. R. Figueira (Eds.), *Multiple criteria decision analysis* (pp. 1269–1298). New York: Springer.

Müller, R., Martinsuo, M., & Blomquist, T. (2008). Project portfolio control and portfolio management performance in different contexts. *Project Management Journal, 39*(3), 28–42.

Niazi, A., Dai, J. S., Balabani, S., & Seneviratne, L. (2006). Product cost estimation: Technique classification and methodology review. *Journal of Manufacturing Science and Engineering, 128*, 563–575.

O'Brien, F. A. (2004). Scenario planning—lessons for practice from teaching and learning. *European Journal of Operational Research, 152*(3), 709–722.

Parvaneh, F., & El-Sayegh, S. M. (2016). Project selection using the combined approach of AHP and LP. *Journal of Financial Management of Property and Construction, 21*(1), 39–53.

Patanakul, P., & Milosevic, D. (2009). The effectiveness in managing a group of multiple projects: Factors of influence and measurement criteria. *International Journal of Project Management, 27*(3), 216–233.

Pérez, F., Gómez, T., Caballero, R., & Liern, V. (2018). Project portfolio selection and planning with fuzzy constraints. *Technological Forecasting and Social Change, 131*, 117–129.

Poh, K. L., Ang, B. W., & Bai, F. (2001). A comparative analysis of R&D project evaluation methods. *R&D Management, 31*(1), 63–75.

Relich, M. (2010a). A decision support system for alternative project choice based on fuzzy neural networks. *Management and Production Engineering Review, 1*(4), 46–54.

Relich, M. (2010b). Assessment of task duration in investment projects. *Management, 14*(2), 136–147.

Relich, M. (2015). Identifying relationships between eco-innovation and product success. In P. Golinska & A. Kawa (Eds.), *Technology management for sustainable production and logistics* (pp. 173–192). Berlin: Springer.

Relich, M. (2016a). A knowledge-based system for new product portfolio selection. In P. Rozewski, D. Novikov, O. Zaikin, & N. Bakhtadze (Eds.), *New frontiers in information and production systems modelling and analysis* (pp. 169–187). Cham: Springer.

Relich, M. (2016b). Portfolio selection of new product projects: A product reliability perspective. *Eksploatacja i Niezawodnosc-Maintenance and Reliability, 18*(4), 613–620.

Relich, M., & Pawlewski, P. (2015). A multi-agent system for selecting portfolio of new product development projects. In *International Conference on Practical Applications of Agents and Multi-Agent Systems* (pp. 102–114). Cham: Springer.

Relich, M., & Pawlewski, P. (2017). A fuzzy weighted average approach for selecting portfolio of new product development projects. *Neurocomputing, 231*, 19–27.

Relich, M., Swic, A., & Gola, A. (2015). A knowledge-based approach to product concept screening. In *12th International Conference on Distributed Computing and Artificial Intelligence* (pp. 341–348). Cham: Springer.

Shirland, L. E., Jesse, R. R., Thompson, R. L., & Iacovou, C. L. (2003). Determining attribute weights using mathematical programming. *Omega, 31*(6), 423–437.

Sun, H., & Ma, T. (2005). A packing-multiple-boxes model for R&D project selection and scheduling. *Technovation, 25*(11), 1355–1361.

Tavana, M., Keramatpour, M., Santos-Arteaga, F. J., & Ghorbaniane, E. (2015). A fuzzy hybrid project portfolio selection method using data envelopment analysis, TOPSIS and integer programming. *Expert Systems with Applications, 42*(22), 8432–8444.

Thieme, R. J., Song, M., & Calantone, R. J. (2000). Artificial neural network decision support systems for new product development project selection. *Journal of Marketing Research, 37*(4), 499–507.

Thiesing, F. M., & Vornberger, O. (1997). Forecasting sales using neural networks. In B. Reusch (Ed.), *Computational Intelligence Theory and Applications* (pp. 321–328). Berlin: Springer.

Tian, Q., Ma, J., & Liu, O. (2002). A hybrid knowledge and model system for R&D project selection. *Expert Systems with Applications, 23*(3), 265–271.

Tietje, O. (2005). Identification of a small reliable and efficient set of consistent scenarios. *European Journal of Operational Research, 162*(2), 418–432.

Tohumcu, Z., & Karasakal, E. (2010). R&D project performance evaluation with multiple and interdependent criteria. *IEEE Transactions on Engineering Management, 57*(4), 620–633.

Ulrich, K. T., & Eppinger, S. D. (2012). *Product design and development* (5th ed.). New York: McGraw-Hill.

Verbano, C., & Nosella, A. (2010). Addressing R&D investment decisions: a cross analysis of R&D project selection methods. *European Journal of Innovation Management, 13*(3), 355–379.

Vidal, L. A., Marle, F., & Bocquet, J. C. (2011). Measuring project complexity using the Analytic Hierarchy Process. *International Journal of Project Management, 29*(6), 718–727.

Wallenius, J., Dyer, J. S., Fishburn, P. C., Steuer, R. E., Zionts, S., & Deb, K. (2008). Multiple criteria decision-making, multiattribute utility theory: Recent accomplishments and what lies ahead. *Management Science, 54*(7), 1336–1349.

Wang, J., & Hwang, W. L. (2007). A fuzzy set approach for R&D portfolio selection using a real options valuation model. *Omega, 35*(3), 247–257.

Wiesemann, W., Kuhn, D., & Rustem, B. (2010). Maximizing the net present value of a project under uncertainty. *European Journal of Operational Research, 202*(2), 356–367.

Yan, H. B., & Ma, T. (2015). A fuzzy group decision-making approach to new product concept screening at the fuzzy front end. *International Journal of Production Research, 53*(13), 4021–4049.

Yang, T., & Hsieh, C. H. (2009). Six-Sigma project selection using national quality award criteria and Delphi fuzzy multiple criteria decision-making method. *Expert Systems with Applications, 36*(4), 7594–7603.

Yazgan, H. R., Boran, S., & Goztepe, K. (2009). An ERP software selection process with using artificial neural network based on analytic network process approach. *Expert Systems with Applications, 36*(5), 9214–9222.

Chapter 2
Model for Formulating Decision Problems Within Portfolio Management

A Framework for Modelling Portfolio Management

Portfolio management mainly refers to the portfolio selection problem that is closely connected with resource-constrained project scheduling problems. Moreover, the selection of the most promising NPD projects for a portfolio requires the evaluation of potential costs, sales, and profit related to a new product. Figure 2.1 illustrates a proposed approach for formulating decision problems in project-oriented companies.

A proposed approach consists of four main parts: data collection, prediction, problem modelling in terms of a CSP, and simulations. In the first part, the data are collected from specifications related to existing products, resources available in a company, and its environment. The proposed approach is dedicated to enterprises that use project management standards, including performance specifications within product development, and techniques needed for project planning and executing. There are the following requirements to ensure the applicability of the proposed approach:

- An enterprise adjusts common project management standards to its needs.
- An enterprise distinguishes phases in the NPD process.
- An enterprise uses standards for specifying tasks in an NPD project and its results.
- An enterprise uses standards for project portfolio management.
- An enterprise measures the success of an NPD project and new product.
- An enterprise measures the impact of NPD projects on corporate financial performance.
- An enterprise uses the results of a financial performance analysis to improve the effectiveness of NPD projects.
- Managers verify the credibility of plans related to an NPD project.
- An enterprise uses the original schedule for monitoring performance in NPD projects.

M. Relich, *Decision Support for Product Development*, Computational Intelligence Methods and Applications,
https://doi.org/10.1007/978-3-030-43897-5_2

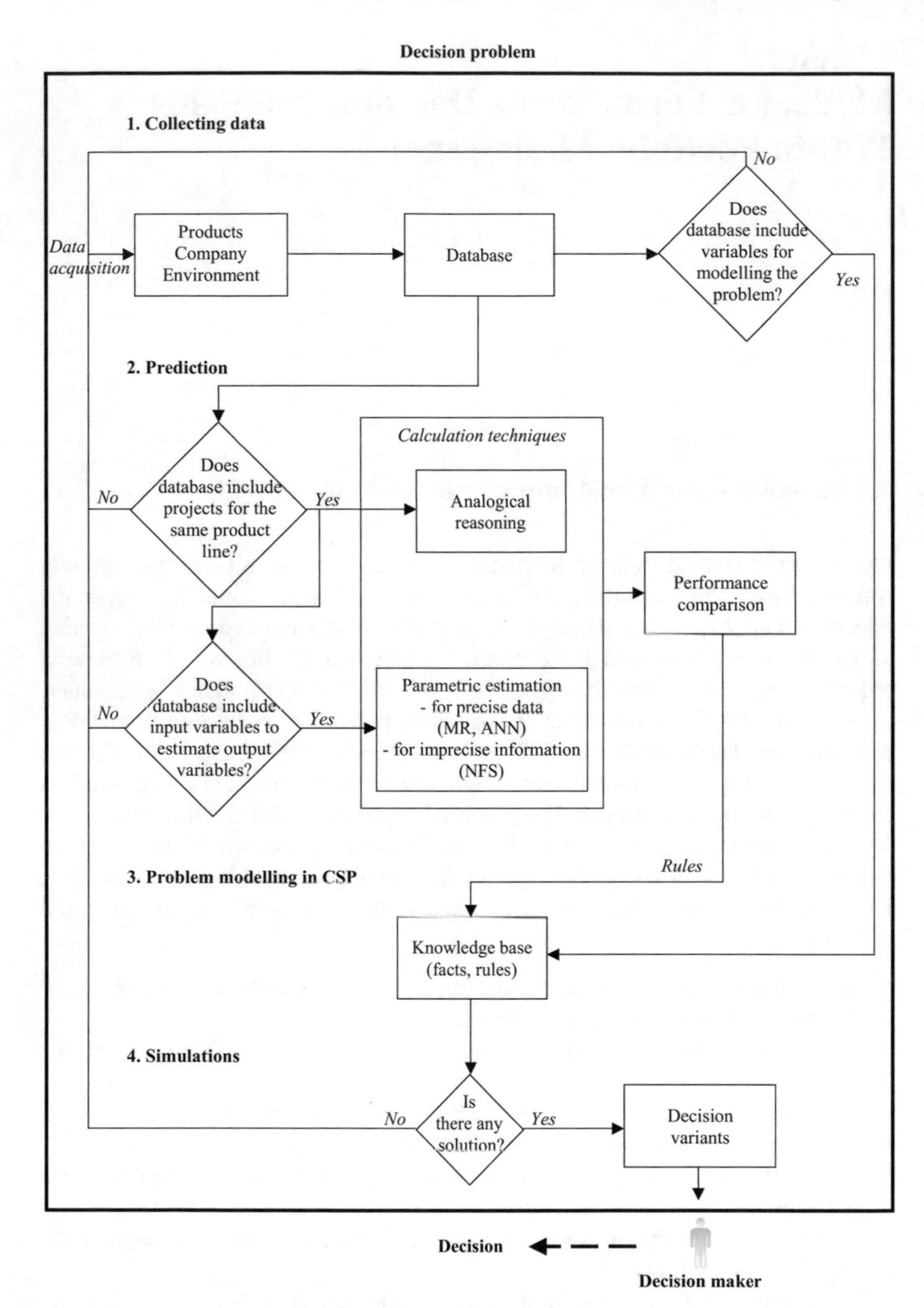

Fig. 2.1 A proposed approach for formulating decision problems

- An enterprise uses the defined procedure to allocate employees to NPD projects.
- An enterprise uses the defined procedure to manage changes in NPD projects according to new business needs.
- An enterprise registers performance and metrics of NPD projects and new products.

The fulfilment of the above requirements for collecting data provides a framework for using the proposed approach.

The second part of the proposed approach is concerned with data analysis and prediction, including a set of calculation techniques to estimate an output variable. If an enterprise has developed NPD projects belonging to the same product line as a new product, then analogical reasoning can be used. Analogical reasoning can help decision-makers obtain information about performance of the past project (e.g. the cost and time of product design) and product after the launch (e.g. unit production cost, marketing cost, sales volume). If the database includes variables that strongly affect an output variable, then parametric estimation can be used. Parametric estimation is concerned with identifying relationships between input and output variables, and depending on notation for numbers, it can use multiple regression (MR) or artificial neural neurons (ANN) for precise numbers and neuro-fuzzy systems (ANFIS) for imprecise information. A parametric estimation method is described in detail in Chap. 3.

The third part refers to modelling the decision problem in terms of a CSP. A knowledge base consists of facts and rules regarding the product, enterprise, and its environment. Relationships identified during data analysis can be also incorporated into the knowledge base. These relationships are produced by an estimation technique that provides the best predictor (i.e. obtains the least error in the testing dataset). Further sections of this chapter illustrate the use of a CSP to specify the portfolio selection problem and resource-constrained project scheduling problem.

The fourth part of the proposed approach is related to simulations that use a CSP framework to search admissible solutions, if they exist. The extent of simulations depends on the decision-maker's requirements regarding the number of involved parameters (e.g. input variables, their domains) and permissible options in a decision support system. These options refer to search prerequisites that ensure the desirable value of an output variable. The aspect of seeking admissible solutions within established constraints is related to the problem specified in an inverse form. In turn, data analysis and prediction described in the second part of the proposed approach can be referred to the problem specified in a forward form. An original methodology dedicated to these two types of problems is presented in Chap. 3.

The results of simulations constitute decision variants that are presented together with consequences for the decision-maker. If there is no solution, then the system should be expanded towards acquiring new data, determining another set of input variables, or variables for modelling the decision problem. Moreover, each completed NPD project expands the database, potentially improving the validity in adaptive data analysis.

A Constraint Satisfaction Problem

A constraint satisfaction problem is composed of a finite set of variables, a set of domains related to these variables, and a finite set of constraints that restricts the values of variables. The solution of a CSP refers to the assignment of a value to each variable satisfying all constraints. A CSP is formulated as follows (Banaszak et al. 2009):

$$\{V,D,C\} \tag{2.1}$$

where V is a finite set of n variables $\{v_1, v_2, \ldots, v_n\}$, D is finite and discrete domains $\{d_1, d_2, \ldots, d_n\}$ from which variables in V take values (where $d_i = \{d_{i1}, d_{i2}, \ldots, d_{ir}\}$), and C is a finite set of constraints $\{c_1, c_2, \ldots, c_m\}$. Each constraint treated as a predicate can be seen as an n-ary relation defined by a Cartesian product $d_1 \times d_2 \times \cdots \times d_n$. An evaluation of variables is a function from variables to domains, v: $V \rightarrow D$. The solution to a CSP is a vector $(d_{1i}, d_{2k}, \ldots, d_{nj})$ such that the values of variables satisfy all constraints C. Generally, constraints can be expressed by arbitrary analytical and/or logical formulas, and they can bind variables with different non-numerical events.

The inference engine consists of two components: constraint propagation and variable distribution. Constraint propagation uses constraints to prune the search space. The aim of propagation techniques is to reach a certain level of consistency in order to accelerate search procedures by reducing the size of the search tree (Banaszak et al. 2009). The values of variables that do not satisfy constraints are removed from their domains during the propagation of constraints (Banaszak and Bocewicz 2011).

Let us consider the following example to illustrate the constraint propagation mechanism. Two variables v_1 and v_2 are related to domains $d_1 \in \{1, \ldots, 5\}$ and $d_2 \in \{1, \ldots, 6\}$, respectively. The constraints are as follows c_1: $v_1 > v_2 + 1$ and c_2: $v_1 \leq v_2$. The constraint c_1 reduces domains to: $d_1 \in \{3, 4, 5\}$ and $d_2 \in \{1, 2, 3\}$. In turn, the constraint c_2 reduces domains to: $d_1 \in \{3\}$ and $d_2 \in \{3\}$. Consequently, there is only one solution that satisfies all constraints. Figure 2.2 illustrates constraint propagation for the considered example.

The resulting set of feasible solutions includes combinations of variable values. The variable distribution phase can start with any variable. If distributions do not result in unique variable domains, the procedure is repeated. Since constraint propagation executes almost immediately, the size of the formulated problem is limited in practical terms in the variable distribution phase, in which the time-consuming backtracking search is used. Consequently, the practical usability of DSS based on constraint programming is related to the development of a variable distribution strategy that could find a feasible solution in an interactive way (Banaszak et al. 2009).

A CSP is solved if a solution tuple exists. Depending on the requirements of an application, CSPs can be classified into the following categories (Tsang 1996):

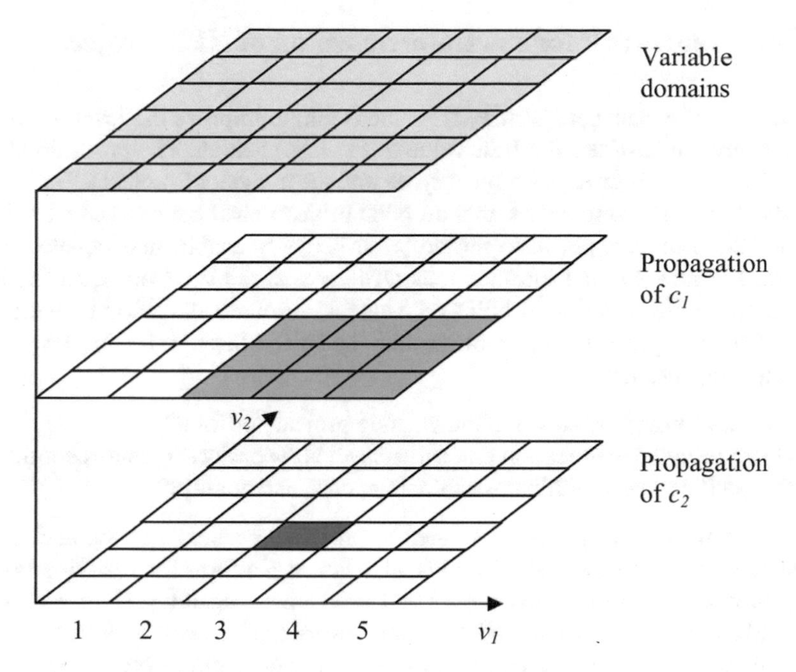

Fig. 2.2 An example of constraint propagation

- A CSP is solved through finding any tuple.
- A CSP is solved through finding all tuples.
- A CSP is solved through finding optimal solutions, where optimality is defied according to some domain knowledge; optimal or near optimal solutions are often required in scheduling.

Constraint satisfaction problems can be solved through the application of a constraint programming (CP). The declarative nature of CP is particularly useful for applications where it is enough to state *what* has to be solved without saying *how* to solve it (Banaszak et al. 2009). As CP uses specific search methods and constraint propagation algorithms, it enables a significant reduction of the search space. Consequently, CP is suitable for modelling complex problems.

A CSP framework can be applied to a wide range of problems, for instance, scheduling (Baptiste et al. 2001; Relich 2011a; Bocewicz et al. 2016), planning (Do and Kambhampati 2001; Relich 2017), manufacturing (Banaszak 2006; Soto et al. 2012), resource allocation (Modi et al. 2001; Relich 2014a), and supply chain problem (Sitek and Wikarek 2008; Chang 2010). In the context of new product development, the CSP paradigm has been used in areas such as design and product configuration (Puget and Van Hentenryck 1998; Yang and Dong 2012), project scheduling (Banaszak and Zaremba 2006; Trojet et al. 2011; Relich 2014b), and project planning (Srivastava et al. 2001; Relich 2011b).

A CSP Framework for Portfolio Selection of NPD Projects

Project portfolio management (PPM) methods aim to improve the level of product success through ensuring the best value to the organisation. The portfolio should contain a balance of different project types and the number of projects. These projects should be limited to ensure that all NPD projects can be resourced effectively. On the other hand, the portfolio should be sufficient to enable an adequate flow of projects and new product introductions (Killen et al. 2008). Consequently, PPM includes portfolio selection of NPD projects and resource allocation between current and new projects. These problems may be solved through finding answers to the following questions:

- Can a new project be added to the existing project portfolio?
- Is there resource reallocation that satisfies all NPD projects (ensures completion of a specific project within desirable time, cost, and quality)?

The specification of problems related to project portfolio selection and project scheduling in terms of a CSP provides an efficient platform for verifying consistency between new production orders and production capability (Banaszak et al. 2009). Moreover, a CP-based modelling framework can be used for development of decision support tools dedicated to project management, being a promising alternative for commercially available tools based on other technologies (e.g. ERP software), the application of which in solving real-life problems is quite limited (Banaszak 2006; Bach et al. 2008). Hence, further considerations develop a CP-based modelling framework towards its usage in selecting an NPD project portfolio.

Figure 2.3 illustrates schematically the constraints, variables, and corresponding relationships that constitute the portfolio selection problem in terms of a CSP. The profit is calculated on the basis of variables that are mainly related to costs and sales.

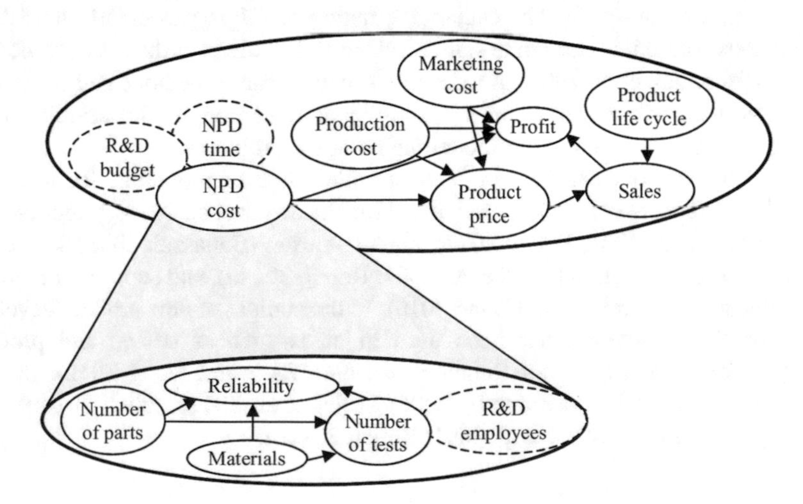

Fig. 2.3 A CSP framework for selecting an NPD project portfolio

In turn, these variables can be estimated with the use of other variables that indirectly affect the profit. For example, the NPD cost is estimated through a parametric model embracing input variables such as the number of parts and tests for a new product, reliability, and materials used.

The proposed CSP framework consists of variables and constraints that can be related to a product, company, and its environment. This structure is further considered for variables specified in the precise form. Nevertheless, there is also research devoted to variable specifications in the imprecise form (Relich 2013, 2015).

Problem Formulation

The problem of project portfolio selection can be specified in terms of a CSP in the following form:

$$\begin{aligned} \mathrm{CSP} = ((&\{P_{it}, R, X, \mathrm{Cma}_i, \mathrm{Ccg}_i, \mathrm{Cpd}_i, \mathrm{Cpt}_i, \mathrm{Cmc}_i, \mathrm{Upc}_i, \mathrm{Up}_i, \mathrm{Sv}_i, \mathrm{Np}_i\}, \\ &\{D_{P_i}, D_R, D_X, D_{\mathrm{Cma}_i}, D_{\mathrm{Ccg}_i}, D_{\mathrm{Cpd}_i}, D_{\mathrm{Cpt}_i}, D_{\mathrm{Cmc}_i}, D_{\mathrm{Upc}_i}, D_{\mathrm{Up}_i}, D_{\mathrm{Sv}_i}, D_{\mathrm{Np}_i}\}), \\ &\{C_1, \ldots, C_{15}\}) \end{aligned} \tag{2.2}$$

The *variables* are defined as follows:

$$P_{it} = \begin{cases} 1 & \text{if the } i\text{ th NPD project is included in the portfolio and starts in period } t \\ 0 & \text{otherwise} \end{cases}$$

for $i = 1, \ldots, I$, where I is the total number of potential NPD projects, and $t = 1, \ldots, T$, when the planning horizon is divided into T periods.

$R = (r_{1,1,1}, r_{2,1,1}, \ldots, r_{1,i,t}, r_{2,i,t}, \ldots, r_{1,I,T}, r_{2,I,T})$ — $r_{1,i,t}$ is the number of employees assigned to the i-th NPD project, and $r_{2,i,t}$ is the amount of financial means needed in the t-th time unit ($t = 0, 1, \ldots, T$)

$X = (x_{1i}, \ldots, x_{22i})$ — is the set of input variables that refers to the i-th product development process:

x_{1i} — the number of potential customers interviewed during market analysis

x_{2i} — the duration of market analysis

x_{3i} — the number of customers' needs recognised during market analysis

x_{4i} — the number of employees involved in market analysis

x_{5i} — the number of substitutive products analysed for elaborating new concepts

x_{6i} the number of ideas (concepts) for a new product
x_{7i} the number of employees involved in elaborating new concepts
x_{8i} the rate of similarity between existing products and a new product
x_{9i} the rate of customers' needs translated into technical specification
x_{10i} the number of components in a new product
x_{11i} the number of employees involved in product design
x_{12i} the duration of product design
x_{13i} the number of adjustments (changes in the product specification) towards reducing the cost and/or time of manufacturing a new product
x_{14i} the number of prototypes for a new product
x_{15i} the number of prototype tests
x_{16i} the number of usage cycles for the first failure
x_{17i} the number of employees involved in testing prototypes
x_{18i} the duration of testing prototypes
x_{19i} the duration of marketing campaign of a new product
x_{20i} the number of potential receivers of marketing campaign
x_{21i} the amount of materials needed to produce a unit of a new product
x_{22i} the time needed to produce a unit of a new product

The set of output variables is as follows:

Cma_i the cost of market analysis of the i-th product
Ccg_i the cost of generating concepts of the i-th product
Cpd_i the cost of designing the i-th product
Cpt_i the cost of testing prototypes of the i-th product
Cmc_i the cost of marketing campaign of the i-th product
Upc_i the unit production cost of the i-th product
Up_i the unit price of the i-th product
Sv_i the sales volume of the i-th product
Np_i the net profit of the i-th product

The above-presented variables are related to the following domains:

D_{Pi} an admissible number of NPD projects in a company
D_{Rk} an admissible number of the k-th resource, $k = 1, 2$
D_{Xj} a range of the domain of the j-th input variable
D_{Cmai} an admissible cost of market analysis of the i-th product
D_{Ccgi} an admissible cost of generating concepts of the i-th product
D_{Cpdi} an admissible cost of designing the i-th product
D_{Cpti} an admissible cost of testing prototypes of the i-th product
D_{Cmci} an admissible cost of marketing campaign of the i-th product
D_{Upci} an admissible unit production cost of the i-th product
D_{Upi} an admissible unit price of the i-th product (limited by the prices of substitute products)
D_{Svi} an admissible sales volume of the i-th product (limited by the market size and competitors' activities)
D_{Npi} an admissible net profit of the i-th product

The following set of *constraints* ensures that each project (if selected) will only start once during the planning horizon:

$$\sum_{t=1}^{T} P_{it} \leq 1 \text{ for } i = 1, \ldots, I \tag{2.3}$$

Appropriate sets of constraints can be established for each limited resource (employees and financial means). The amount of resources available to carry out a set of projects may change over time. For example, if the planning horizon is divided into T planning periods, and the maximum allowed cost for all projects during period k should not exceed a certain amount (AF_k), then the set of constraints is as follows:

$$\sum_{i=1}^{I} \sum_{t=1}^{k} C_{i,k+1-t} P_{it} \leq \mathrm{AF}_k \text{ for } k = 1, \ldots, T \tag{2.4}$$

where AF_k is the total financial means available in period k and $C_{i,k+1-t}$ is the financial means required by the i-th project in the k-th period. Note that if the i-th project starts in the t-th period, it is in its $(k - t + 1)$th period in the k-th period, and so will need $C_{i,k+1-t}$ financial means. This constraint ensures that each project (if started) should continue to completion within the planning horizon.

The following constraint refers to the issue that all of the selected projects should be completed within the planning horizon:

$$\sum_{t=1}^{T} tP_{it} + D_i \leq T + 1 \text{ for } i = 1, \ldots, I \tag{2.5}$$

where D_i is the duration of the i-th project.

Different levels of resources allocated to a project can lead to the acceleration or slowdown of project completion. Significant changes in these levels can result in large disparities between planned and actual values of project performance. The following constraint is needed to ensure that only one version of the project will be selected:

$$\sum_{i \in U_v} \sum_{t=1}^{T} P_{it} \leq 1 \tag{2.6}$$

where U_v is the set that contains different versions of an individual project.

The project portfolio can include projects that are significant from the strategic point of view and that should be definitely continued. As projects compete for scarce resources, it seems to be important to take the issue of mandatory and ongoing projects into consideration. The following constraints ensure the placement of these projects in the portfolio:

$$\sum_{t=1}^{T} P_{it} = 1 \text{ for } i \in U_m \tag{2.7}$$

$$P_{il} = 1 \text{ for } i \in U_0 \tag{2.8}$$

where U_m is the set of mandatory projects, and U_0 is the set of ongoing projects that should be continued.

There is also a need to determine the impact of removing certain ongoing projects from the portfolio:

$$\sum_{t=1}^{T} P_{it} = 0 \text{ for } i \in U_d \tag{2.9}$$

where U_d is the set of ongoing projects that should be removed from the portfolio.

Another issue that should be included in the portfolio selection problem is interdependence between projects. For example, if project P_1 is dependent on project P_2, then project P_2 must be included if project P_1 is selected in the portfolio. Nevertheless, project P_2 can be selected in the portfolio even if project P_1 is removed. For example, a product from one product line might be dependent on a product from another product line. These types of interdependencies between projects can be described in the following constraints:

$$\sum_{t=1}^{T} P_{it} \geq \sum_{t=1}^{T} P_{lt} \tag{2.10}$$

$$\sum_{t=1}^{T} tP_{lt} + (T+1) \cdot \left(1 - \sum_{t=1}^{T} P_{lt}\right) - \sum_{t=1}^{T} tP_{it} \geq D_i \sum_{t=1}^{T} P_{it} \tag{2.11}$$

for $i \in \text{PP}_l$, where PP_l is the set of precursor projects for a particular project l; $l = 1, \ldots, L$.

Constraint (2.10) ensures the selection of its precursor projects, once a project is selected, and constraint (2.11) ensures that all of the precursor projects will be finished before the successor project starts.

Interdependence can be also considered from the perspective of mutual exclusiveness. For example, the selection of an NPD project from the specific product line excludes the selection of another NPD project from the same product line.

$$\sum_{i \in U_p} \sum_{t=1}^{T} P_{it} \leq 1 \text{for } p = 1, \ldots, \text{PP} \tag{2.12}$$

where PP are sets of mutually exclusive projects, and U_p is the p-th set of such projects.

Another issue is the selection of a portfolio of NPD projects that balances the overall development risk. Typically, more risky projects are related to greater expected benefits (if completed successfully). For example, a balanced portfolio includes a small investment in potentially high-risk and high-benefit projects and more investment in low-risk and low-benefit projects. A portfolio of projects with different risks allows an enterprise to achieve the desirable results. The following constraints ensure consideration of investment in high-risk and long-term projects:

$$\sum_{i \in U_h} \left(\text{TC}_i \sum_{t=1}^{T} P_{it} \right) \leq \text{PHR} \cdot \left(\sum_{k=1}^{n} \left(\text{TC}_k \sum_{t=1}^{T} P_{kt} \right) \right) \tag{2.13}$$

$$\sum_{i \in U_l} \left(\text{TC}_i \sum_{t=1}^{T} P_{it} \right) \leq \text{PLT} \cdot \left(\sum_{k=1}^{n} \left(\text{TC}_k \sum_{t=1}^{T} P_{kt} \right) \right) \tag{2.14}$$

where TC_i is the total cost of executing the i-th project, U_{h} and U_{l} are the sets of high-risk and long-term projects, and PHR and PLT are the maximum desirable level of investment in high-risk and long-term projects, respectively.

The total cost (TC_i) of product development and launch of the i-th product consists of the cost of market analysis (Cma_i), concept generation (Ccg_i), product design (Cpd_i), prototype tests (Cpt_i), and marketing campaign (Cmc_i):

$$\text{TC}_i = \text{Cma}_i + \text{Ccg}_i + \text{Cpd}_i + \text{Cpt}_i + \text{Cmc}_i \tag{2.15}$$

The predicted net profit (Np_{it}) for the i-th product in the t-th period is calculated as follows:

$$\text{Np}_{it} = \text{Sv}_{it} \cdot (\text{Up}_{it} - \text{Upc}_{it}) - \text{TC}_i \tag{2.16}$$

The profitability evaluation of an NPD project over a longer time period imposes involvement of discounted cash flows (e.g. used in the NPV method), in which net present values referring to the predicted inflows and outflows are adjusted to the interest rate in different time periods.

The *objective function* is as follows:

$$\max Z = \sum_{i=1}^{N} \sum_{j=1}^{T} a_i P_{ij} \tag{2.17}$$

where Z is the value function that is maximised, and a_i is the score (in a qualitative approach) or NPV (in a quantitative approach) of project i.

A CSP can be considered as a knowledge base that is a platform for query formulation and obtaining answers. The solution of the project portfolio selection problem is related to seeking answers to the following questions:

1. Is there any NPD project portfolio that fulfils the assumed constraints, and if yes, which NPD projects constitute this portfolio?
2. What values should be assigned to the decision variables to fulfil the assumed constraints?

The first question is referred to a forward approach and the second to an inverse approach. A forward approach refers to problem solution, in which the values of the decision variables determine the values of an objective function. In turn, an inverse approach is related to problem solution, in which a set of the values of the decision variables is sought to achieve the desirable value of an objective function.

An Example of Project Portfolio Selection

An example consists of two cases that refer to the project portfolio selection problem stated in the forward and inverse form.

Case 1: Project Portfolio Selection in the Forward Form

Let us assume that a set of potential new products consists of five NPD projects $\{P_1, \ldots, P_5\}$. The selected portfolio should be completed in a time span of 30 weeks and within the R&D budget of 300 thousand €. Moreover, the company can delegate seven employees to R&D teams that will be responsible for developing new products. Top management tends to form two project teams that will involve all available R&D employees in the NPD process. Table 2.1 presents evaluation criteria for selecting an NPD project portfolio.

Table 2.1 Evaluation criteria for project portfolio selection

Criteria	P_1	P_2	P_3	P_4	P_5
The number of R&D team members—R	4	3	4	3	3
Time of product development (in weeks)—D	30	24	32	28	26
Cost of product development (in thousand €)—TC	155.7	142.9	161.1	151.6	147.3
Net profit (in million €)—Np	5.3	4.8	5.7	5.2	5.0

The objective function is the total discounted net profit in the expected product life cycle. The optimal portfolio includes projects P_4 and P_5, with the expected total cost of development equalling 298.9 thousand €, and the total net profit—10.2 million €.

Case 2: Project Portfolio Selection in the Inverse Form

Let us assume that the decision-maker is interested in developing two NPD projects to reach the net profit greater than 10.2 millions €, and not exceed the R&D budget (300 thousand €). The set of input variables includes the number of parts (X_{10}), the number of prototype tests (X_{15}), and the amount of materials (X_{21}) related to a new product. These variables impact the cost of product development and indirectly net profit (see Fig. 2.3). The relationships between input variables and the cost of product development are identified with the use of parametric estimation techniques (e.g. based on computational intelligence).

The use of an inverse approach requires the specification of domains for input variables that are as follows: P_3 $\{X_{10} = 36\#37, X_{15} = 12\#13, X_{21} = 5\#6\}$, and P_4 $\{X_{10} = 32\#33, X_{15} = 11\#12, X_{21} = 4\#5\}$. Table 2.2 presents possible variants for different values of decision variables within the specified constraints for projects P_3 and P_4.

The specification of the portfolio selection problem in the inverse form allows the decision-maker to obtain information about possible variants of the R&D portfolio, indicating potential directions of creating optimal portfolio. On the other hand, the potential large search space imposes the use of search techniques that can reduce the search space, and the relevant time needed to find possible solutions. One of these techniques is constraint programming that can effectively solve the portfolio selection problem stated in terms of a CSP.

Table 2.2 Variants of the NPD portfolio for the desirable NPD cost

	Values of decision variables	ΣTC
Variant 1	P_3 $\{X_{10} = 36, X_{15} = 12, X_{21} = 5\}$, P_4 $\{X_{10} = 32, X_{15} = 11, X_{21} = 4\}$	298,090
Variant 2	P_3 $\{X_{10} = 37, X_{15} = 12, X_{21} = 5\}$, P_4 $\{X_{10} = 32, X_{15} = 11, X_{21} = 4\}$	299,090
Variant 3	P_3 $\{X_{10} = 36, X_{15} = 12, X_{21} = 5\}$, P_4 $\{X_{10} = 33, X_{15} = 11, X_{21} = 4\}$	298,840
Variant 4	P_3 $\{X_{10} = 37, X_{15} = 12, X_{21} = 5\}$, P_4 $\{X_{10} = 33, X_{15} = 11, X_{21} = 4\}$	299,840
Variant 5	P_3 $\{X_{10} = 36, X_{15} = 12, X_{21} = 5\}$, P_4 $\{X_{10} = 32, X_{15} = 12, X_{21} = 4\}$	299,450

A CSP Framework for Scheduling NPD Projects

The resource-constrained project scheduling problem (RCPSP) consists of precedence constraints and resource constraints for scheduling the project activities (Herroelen et al. 1998). The resource-constrained multi-project scheduling problem (RCMPSP) is one of the variants of RCPSP. RCMPSP has been variously considered in the literature on project management. Mingozzi et al. (1998) present exact optimisation approaches for scheduling single and multiple projects. Ghasemzadeh et al. (1999) provide a zero-one model for project portfolio selection and scheduling. In turn, Gonçalves et al. (2008), Beşikci et al. (2015), and Yassine et al. (2017) present the use of genetic algorithms for solving RCMPSP. Kyriakidis et al. (2012) provide a mixed-integer linear programming (MILP) model for the deterministic single- and multi-mode RCPSP, whereas Dash et al. (2018) propose a MILP model for planning the product portfolio and launch timings under resource constraints. Kopanos et al. (2014) and Kreter et al. (2016) present models and solution procedures for solving RCPSP with temporal constraints using continuous- and discrete-time mathematical formulations. In turn, Relich (2016) presents a knowledge-based system for solving RCMPSP specified in terms of a CSP and dedicated to NPD projects.

Product development depends on factors related to the business environment and company resources. For example, the customers' requirements and comments may suggest directions of new product design that is developed within the limitations related to the company (e.g. the R&D budget, the number of R&D employees) and its environment (e.g. cost of materials, product safety regulations, available technology). The constraints can link and limit the decision variables, e.g. the number of R&D employees impacts the completion time of an NPD project. The NPD model specified in the declarative representation (as variables and constraints) enables the formulation of project scheduling problem in terms of a CSP, allowing implementation of the problem in the CP environment.

Problem Formulation

Let us assume that a project portfolio includes I projects: $P = \{P_1, P_2, \ldots, P_I\}$, and a single project P_i consists of J activities: $P_i = \{A_{i,1}, \ldots, A_{i,j}, \ldots, A_{i,J}\}$. It is assumed that an activity is indivisible, and it can start only if the required amount of resources is available at the moments given by $\mathrm{Tp}_{i,j}$ and after completion of previous activities. The resource can be allotted or released only after completion of the activity that requires this resource. The project P_i is described as an activity-on-node network, where nodes represent the activities, and the arcs determine the precedence constraints between activities.

The scheduling problem in terms of the CSP can be specified as follows:

$$\begin{aligned} \text{CSP} = ((&\{R, P_i, A_{i,j}, s_{i,j}, t_{i,j}, \text{Tp}_{i,j}, \text{Tz}_{i,j}, \text{Dp}_{i,j}\}, \\ &\{D_R, D_{P_i}, D_{A_i}, D_{s_i}, D_{t_i}, D_{\text{Tp}_i}, D_{\text{Tz}_i}, D_{\text{Dp}_i}\}), \\ &\{C_1, \ldots, C_7\}) \end{aligned} \tag{2.18}$$

where:

$R = (r_{1,1}, r_{2,1}, r_{3,1}, \ldots, r_{1,t}, r_{2,t}, r_{3,t}, \ldots, r_{1,T}, r_{2,T}, r_{3,T})$ — $r_{1,t}$ is the number of employees assigned to an NPD project, $r_{2,t}$ is the financial means, and $r_{3,t}$ is the amount of materials needed in the t-th time unit ($t = 0, 1, \ldots, T$)

P_i — i-th project

$A_{i,j}$ — j-th activity of the i-th project, in which $A_{i,j} = \{s_{i,j}, t_{i,j}, \text{Tp}_{i,j}, \text{Tz}_{i,j}, \text{Dp}_{i,j}\}$

$s_{i,j}$ — the start time of the activity $A_{i,j}$, i.e. the time counted from the beginning of the time horizon T

$t_{i,j}$ — the duration of the activity $A_{i,j}$

$\text{Tp}_{i,j} = (\text{tp}_{i,j,1}, \text{tp}_{i,j,2}, \text{tp}_{i,j,3}, \ldots, \text{tp}_{I,J,3})$ — the sequence of moments, in which the k-th resource is allocated to the activity $A_{i,j}$; $\text{tp}_{i,j,k}$ is the time counted since the moment $s_{i,j}$; the k-th resource is allocated to the activity $A_{i,j}$ during its execution period: $0 \leq tp_{i,j,k} < t_{i,j}$, $k = 1, 2, 3$;

$\text{Tz}_{i,j} = (\text{tz}_{i,j,1}, \text{tz}_{i,j,2}, \text{tz}_{i,j,3}, \ldots, \text{tz}_{I,J,3})$ — the sequence of moments, in which the activity $A_{i,j}$ releases the resources; $\text{tz}_{i,j,k}$ is the time counted since the moment $s_{i,j}$; the k-th resource is released by the activity $A_{i,j}$ during its execution period: $0 < \text{tz}_{i,j,k} \leq t_{i,j}$, $\text{tp}_{i,j,k} < \text{tz}_{i,j,k}$

$\text{Dp}_{i,j} = (\text{dp}_{i,j,1}, \text{dp}_{i,j,2}, \text{dp}_{i,j,3}, \ldots, \text{dp}_{I,J,3})$ — the sequence of the k-th resource allocations to the activity $A_{i,j}$; $dp_{i,j,k}$ is a number of the k-th resource allocation to the activity $A_{i,j}$

D_R a set of an admissible amount of the k-th resource

D_{Pi} a set of an admissible number of NPD projects in a company

D_{Ai} a set of an admissible number of activities in the i-th project

D_{si} a set of admissible start times of activity $A_{i,j}$ in the i-th project

D_{ti} a set of admissible durations of activity $A_{i,j}$ in the i-th project

$D_{\mathrm{Tp}i}$ a set of admissible allocation moments to activity $A_{i,j}$ for the k-th resource in amount of $\mathrm{dp}_{i,j,k}$ in the i-th project

$D_{\mathrm{Tz}i}$ a set of admissible release moments by activity $A_{i,j}$ for the k-th resource in amount of $\mathrm{dp}_{i,j,k}$ in the i-th project

$D_{\mathrm{Dp}i}$ a set of an admissible number of required resources by the activity $A_{i,j}$ in the i-th project

C_1 the maximal number of R&D employees available in the t-th time unit

C_2 the maximal financial means available in the t-th time unit

C_3 the maximal amount of materials available in the t-th time unit

C_4 the i-th project completion time:

$$\forall s_{i,j} \forall t_{i,j} \left(s_{i,j} + t_{i,j} \leq T \right) \tag{2.19}$$

C_5 precedence constraints:

- The n-th activity follows the j-th one:

$$s_{i,j} + t_{i,j} \leq s_{i,n} \tag{2.20}$$

- The n-th activity follows other activities:

$$\begin{gathered} s_{i,j} + t_{i,j} \leq s_{i,n} \\ s_{i,j+1} + t_{i,j+1} \leq s_{i,n} \\ \ldots \\ s_{i,j+n} + t_{i,j+m} \leq s_{i,n} \end{gathered} \tag{2.21}$$

- The n-th activity is followed by other activities:

$$
\begin{gathered}
s_{i,n} + t_{i,n} \leq s_{i,j} \\
s_{i,n} + t_{i,n} \leq s_{i,j+1} \\
\ldots \\
s_{i,n} + t_{i,n} \leq s_{i,j+m}
\end{gathered}
\tag{2.22}
$$

C_6 the total number of employees $r_{1,t}$ in the company in the t-th time unit cannot be less than the number of employees $r_{1,t,i}$ for the project portfolio:

$$r_{1,t,i} \; '' \; r_{1,t} \tag{2.23}$$

C_7 the total financial means $r_{2,t}$ in the company in the t-th time unit cannot be less than the financial means for the project portfolio $r_{2,t,i}$:

$$r_{2,t,i} \; '' \; r_{2,t} \tag{2.24}$$

C_8 the total amount of materials $r_{3,t}$ in the company in the t-th time unit cannot be less than the amount of materials for the project portfolio $r_{3,t,i}$:

$$r_{3,t,i} \; '' \; r_{3,t} \tag{2.25}$$

A CSP can be considered as a knowledge base that is a platform for query formulation and obtaining answers. A distinction of decision variables that are embedded in the knowledge base as an input–output variable permits the formulation of two classes of standard routine questions that concern two different problems with respect to resources:

- What is the minimal cost and time of project completion under resource constraints?
- What resources and in which amount are necessary to complete the project within the desirable deadline and budget?

The first question can be referred to a forward approach and the second to an inverse approach.

An Example of Project Scheduling

An example consists of two NPD projects that should be scheduled according to available R&D employees. The first project (P_1) is concerned with redesigning a car cockpit towards incorporating green materials in cockpit design. Table 2.3 presents the description of activities in project P_1.

Table 2.3 Activities related to project P_1

Activity	Description	Duration (months)	Predecessor(s)
$A_{1,1}$	Designing research plan	1	–
$A_{1,2}$	Assessing legal regulations, patents, and environmental issues	2	$A_{1,1}$
$A_{1,3}$	Searching for material suppliers	1	$A_{1,2}$
$A_{1,4}$	Searching for experts to evaluate the attractiveness of a new product	1	$A_{1,2}$
$A_{1,5}$	Testing materials	4	$A_{1,3}$
$A_{1,6}$	Designing cockpit including new materials	6	$A_{1,3}$
$A_{1,7}$	Evaluating the attractiveness of a new product	1	$A_{1,4}$, $A_{1,6}$
$A_{1,8}$	Drafting project specifications	2	$A_{1,5}$, $A_{1,6}$, $A_{1,7}$
$A_{1,9}$	Verifying test series of new products	2	$A_{1,8}$

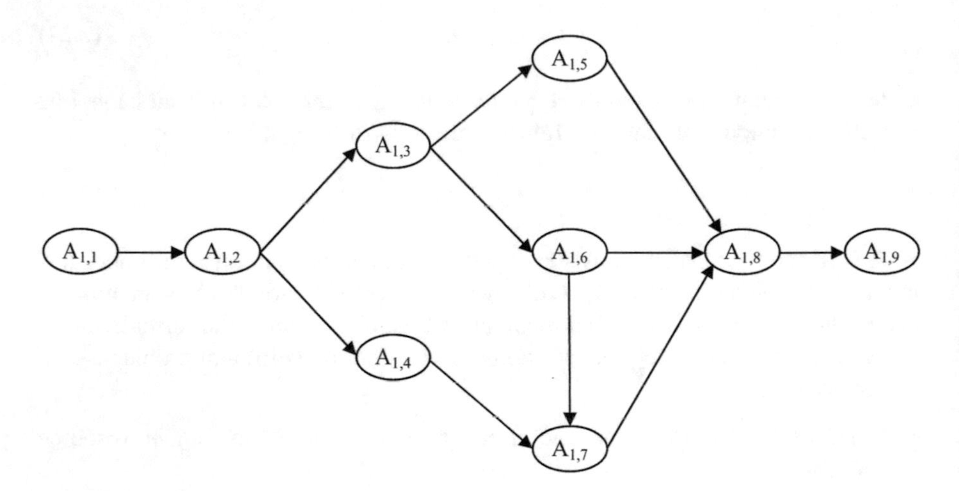

Fig. 2.4 Network diagram for project P_1

The network diagram for activities concerning project P_1 is presented in Fig. 2.4. The planned project completion time reaches 15 months.

The second project (P_2) is concerned with designing new headlights and other connected components. The description of activities in project P_2 is presented in Table 2.4.

The network diagram for activities related to project P_2 is illustrated in Fig. 2.5. The planned project completion time reaches 15 months.

Activities related to the above-presented projects ($P = \{P_1, P_2\}$) can be described in the following sequences: $P_1 = \{A_{1,1}, \ldots, A_{1,9}\}$ and $P_2 = \{A_{2,1}, \ldots, A_{2,8}\}$. The deadline of the project portfolio completion equals 15 months and the R&D

Table 2.4 Activities related to project P_2

Activity	Description	Duration (months)	Predecessor(s)
$A_{2,1}$	Research design	1	–
$A_{2,2}$	Initial financial and risk analysis	1	$A_{2,1}$
$A_{2,3}$	R&D for new components related to headlights	4	$A_{2,2}$
$A_{2,4}$	R&D for a new design of headlights	6	$A_{2,2}$
$A_{2,5}$	R&D for adjusting a car bonnet to headlights	2	$A_{2,2}$, $A_{2,4}$
$A_{2,6}$	Integration tests	3	$A_{2,3}$, $A_{2,4}$, $A_{2,5}$
$A_{2,7}$	Drafting project specifications	2	$A_{2,3}$, $A_{2,4}$, $A_{2,5}$
$A_{2,8}$	Analysis of production test series	2	$A_{2,6}$, $A_{2,7}$

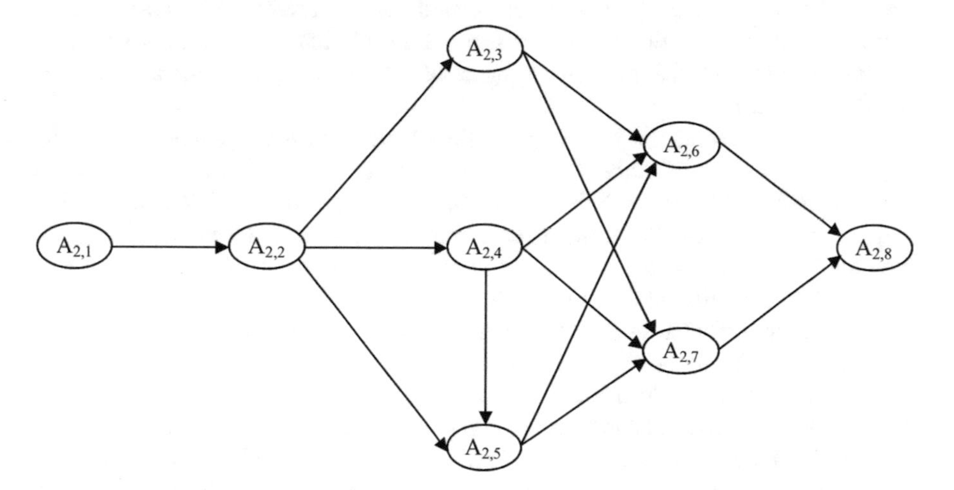

Fig. 2.5 Network diagram for project P_2

Table 2.5 The number of R&D employees for project activities

	A_1	A_2	A_3	A_4	A_5	A_6	A_7	A_8	A_9
$R_{1,1}$	1	1	1	1	2	2	1	1	1
$R_{1,2}$	1	1	2	2	2	1	1	1	–

budget reaches 250 thousand €. The number of R&D employees assigned to project activities is presented in Table 2.5. The sequences of activity duration for the above-presented projects are as follows: $T_1 = (1, 2, 1, 1, 4, 6, 1, 2, 2)$ and $T_2 = (1, 1, 4, 6, 2, 3, 2, 2)$.

The employees can perform independently and simultaneously project activities at the monthly cost of 5 thousand €. The reduction of activity duration is proportional to the number of employees. For example, if activity duration is planned for 4 months and 2 employees are assigned to its execution, then its duration will shorten to 2 months. The project teams related to project P_1 and P_2 can include maximal 4 employees.

A Forward Approach

Problem solution is related to the determination of starting times of project portfolio activities s_{ij}, and the allocation of resources to the activities $dp_{i,j,k}$. The following sequences have been specified for the above-presented project portfolio: $S_1 = (s_{1,1}, \ldots, s_{1,9})$, $S_2 = (s_{2,1}, \ldots, s_{2,8})$ and $Dp_1 = (dp_{1,1,1}, \ldots, dp_{1,9,1})$, $Dp_2 = (dp_{2,1,1}, \ldots, dp_{2,8,1})$. The scheduling problem in the forward approach refers to the answer to the following question: is there resource allocation such that it fulfils all project constraints, and if yes, what are starting times of activities and resource allocation at the minimal cost?

Figure 2.6 presents the project portfolio schedule for the first admissible solution. The sequences of activity starting time are as follows (all activities start as soon as possible): $S_1 = (0, 1, 3, 3, 4, 6, 9, 10, 12)$ and $S_2 = (0, 1, 2, 4, 7, 8, 8, 11)$. The sequences of resource allocation in the forward approach are presented in Table 2.5. The total planned cost reaches 205 thousand €, whereas the planned completion time for longer project (P_1) is 14 months.

Let us assume that after 5 months activity $A_{1,5}$ is behind schedule. Re-estimation indicates that the duration of activities $A_{1,5}$ and $A_{1,6}$ will be extended to 3 and 4 months, respectively. As a result, the total planned cost related to the project portfolio reaches 225 thousand €, whereas completion time is 16 months, extending the planned deadline. Thus, the considered problem is reformulated towards stating it in an inverse approach that aims to search values of decision variables (e.g. reallocation of resources), ensuring project completion within the desirable budget and deadline.

An Inverse Approach

An inverse approach uses the same activity networks and constraints regarding the deadline and budget of the project portfolio as in the forward approach. Problem solution relies on finding an answer to the following question: is there any resource allocation, and if yes, what amount of resources ensures the project portfolio completion within the R&D budget and deadline? The response to this question requires determining the sequences regarding the number of R&D employees assigned to projects (Dp) and starting times of activities (S).

Let us assume that the decision-maker considers the assignment of one additional employee to activities $A_{1,5}$ and/or $A_{1,6}$ in the project P_1 to verify the possibility of reducing the project completion time. There are eight possible variants of the

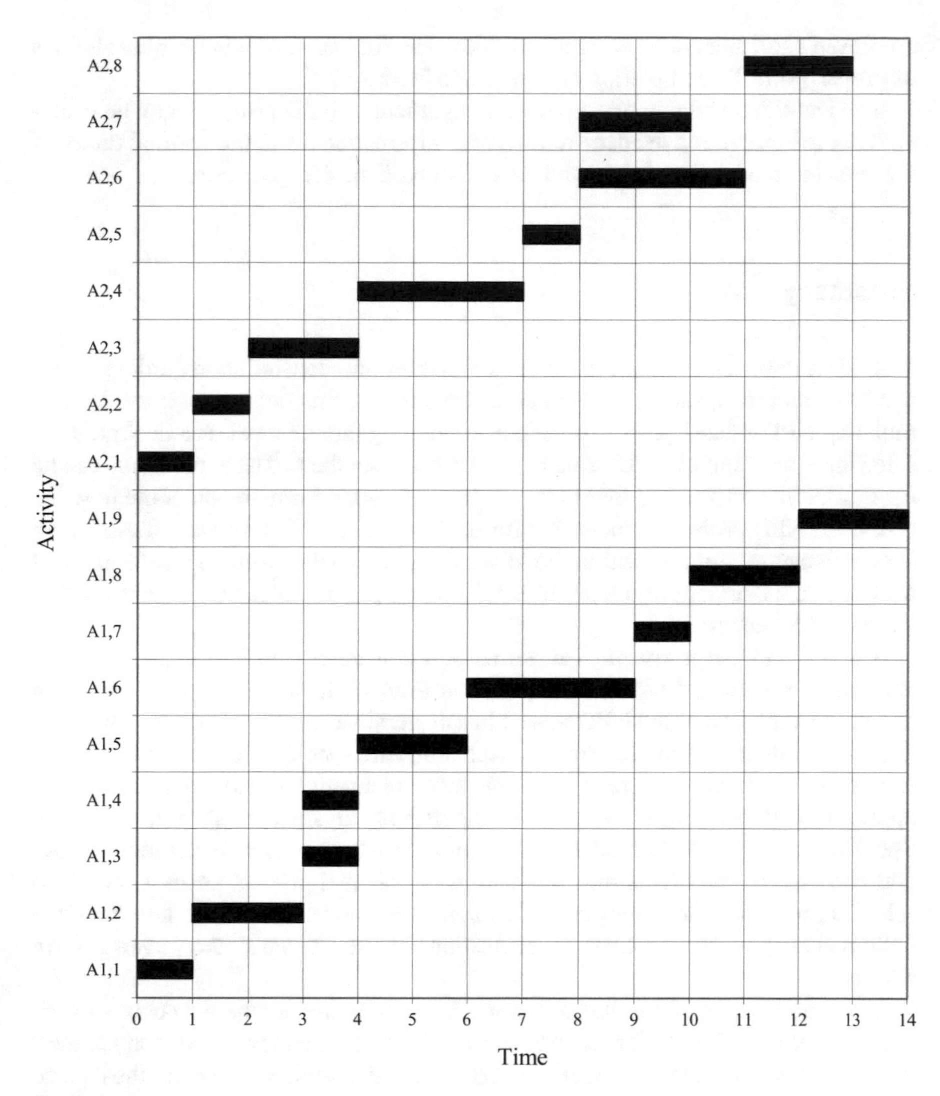

Fig. 2.6 Project portfolio schedule

Table 2.6 Possible variants for project completion

	Employee assignment	Total time	Total costs
Variant 1	$Dp_1 = (1, 1, 1, 1, 3, 3, 1, 1, 1)$ $Dp_2 = (1, 1, 2, 1, 2, 1, 1, 1)$	15 15	240
…	…	…	…
Variant 8	$Dp_1 = (1, 1, 1, 1, 2, 3, 1, 1, 1)$ $Dp_2 = (1, 1, 2, 2, 1, 1, 1, 1)$	15 15	245

employee's assignment to project activities. The first and last admissible solutions of project portfolio completion are presented in Table 2.6.

The identified variants of employee assignment to NPD projects may be evaluated towards providing the decision-makers information about the optimal trade-off between the total planned time and cost of an NPD project portfolio.

Summary

A product development model contains variables and constraints regarding a new product, enterprise, and its environment. The proposed model is a platform for formulating NPD-related problems such as evaluating the potential of a new product, selecting a portfolio of NPD projects, and scheduling them. These problems can be formulated in a forward or inverse form. In the forward form, the outcome (e.g. the cost of an NPD project) is identified for declared values of input variables. If there is no satisfactory solution within the forward form, then the problem is reformulated into an inverse form, in which possible values of input variables are sought to ensure the desirable outcome.

The NPD model is specified in terms of a constraint satisfaction problem that consists of variables, their domains, and constraints. The use of the CSP paradigm enables formulation of the NPD model in a declarative representation, in which the desired results are specified without explaining the specific algorithms needed to achieve these results. Current NPD models are mainly based on a procedural approach that uses a set of defined algorithms to solve a specific problem. Specification of the NPD model in terms of the CSP enables consideration of several NPD-related problems (e.g. evaluation of potential product success, portfolio selection, resource allocation) within a single NPD model. Moreover, the use of the CSP makes possible problem specification in the forward and inverse form alternately.

The search space of possible solutions depends on the number of decision variables and their domains. The increment in decision variables and/or ranges their domains causes that effective identification of all admissible solutions in the inverse approach is a challenging task. To reduce the search space and solve the CSP effectively, the use of constraint programming is proposed. Moreover, the application of CP enables time reduction of finding solutions and consequently improves interactive properties of a decision support system.

To sum up, the declarative approach facilitates the NPD model updating and enables the specification of NPD-related problems in the forward and inverse form. It can be also used as a pertinent platform to model specification and problem formulation in other managerial applications. Nevertheless, the applicability of the proposed approach depends on some requirements related to the project maturity of an organisation. The proposed approach is dedicated to enterprises that use project management standards, including performance specifications within product development, and techniques needed for project planning and executing.

References

Bach, I., Bocewicz, G., & Banaszak, Z. (2008). Constraint programming approach to time-window and multiresource-constrained projects portfolio prototyping. In *New frontiers in applied artificial intelligence* (pp. 767–776). Berlin: Springer.

Banaszak, Z. (2006). CP-based decision support for project driven manufacturing. In *Perspectives in modern project scheduling* (pp. 409–437). Boston, MA: Springer.

Banaszak, Z., & Bocewicz, G. (2011). *Decision support driven models and algorithms of artificial intelligence*. Warsaw: Warsaw University of Technology.

Banaszak, Z., & Zaremba, M. B. (2006). Project-driven planning and scheduling support for virtual manufacturing. *Journal of Intelligent Manufacturing, 17*(6), 641–651.

Banaszak, Z., Zaremba, M., & Muszyński, W. (2009). Constraint programming for project-driven manufacturing. *International Journal of Production Economics, 120*, 463–475.

Baptiste, P., Le Pape, C., & Nuijten, W. (2001). *Constraint-based scheduling: Applying constraint programming to scheduling problems*. Norwell: Kluwer Academic.

Beşikci, U., Bilge, Ü., & Ulusoy, G. (2015). Multi-mode resource constrained multi-project scheduling and resource portfolio problem. *European Journal of Operational Research, 240*(1), 22–31.

Bocewicz, G., Nielsen, I. E., & Banaszak, Z. (2016). Production flows scheduling subject to fuzzy processing time constraints. *International Journal of Computer Integrated Manufacturing, 29*, 1105–1127.

Chang, Y. H. (2010). Adopting co-evolution and constraint-satisfaction concept on genetic algorithms to solve supply chain network design problems. *Expert Systems with Applications, 37*(10), 6919–6930.

Dash, B., Gajanand, M. S., & Narendran, T. T. (2018). A model for planning the product portfolio and launch timings under resource constraints. *International Journal of Production Research, 56*, 5081–5103.

Do, M., & Kambhampati, S. (2001). Planning as constraint satisfaction: Solving the planning graph by compiling it into CSP. *Artificial Intelligence, 132*, 151–182.

Ghasemzadeh, F., Archer, N., & Iyogun, P. (1999). A zero-one model for project portfolio selection and scheduling. *Journal of the Operational Research Society, 50*(7), 745–755.

Gonçalves, J. F., Mendes, J. J., & Resende, M. G. (2008). A genetic algorithm for the resource constrained multi-project scheduling problem. *European Journal of Operational Research, 189*(3), 1171–1190.

Herroelen, W., De Reyck, B., & Demeulemeester, E. (1998). Resource-constrained project scheduling: a survey of recent developments. *Computers & Operations Research, 25*(4), 279–302.

Killen, C. P., Hunt, R. A., & Kleinschmidt, E. J. (2008). Project portfolio management for product innovation. *International Journal of Quality & Reliability Management, 25*(1), 24–38.

Kopanos, G., Kyriakidis, T., & Georgiadis, M. (2014). New continuous-time and discrete-time mathematical formulations for resource-constrained project scheduling problems. *Computers & Chemical Engineering, 68*, 96–106.

Kreter, S., Rieck, J., & Zimmermann, J. (2016). Models and solution procedures for the resource-constrained project scheduling problem with general temporal constraints and calendars. *European Journal of Operational Research, 251*(2), 387–403.

Kyriakidis, T., Kopanos, G., & Georgiadis, M. (2012). MILP formulations for single- and multi-mode resource-constrained project scheduling problems. *Computers and Chemical Engineering, 36*, 369–385.

Mingozzi, A., Maniezzo, V., Ricciardelli, S., & Bianco, L. (1998). An exact algorithm for the resource-constrained project scheduling problem based on a new mathematical formulation. *Management Science, 44*(5), 714–729.

Modi, P. J., Jung, H., Tambe, M., Shen, W. M., & Kulkarni, S. (2001). A dynamic distributed constraint satisfaction approach to resource allocation. In *Principles and practice of constraint programming* (pp. 685–700). Berlin: Springer.

Puget, J. F., & Van Hentenryck, P. (1998). A constraint satisfaction approach to a circuit design problem. *Journal of Global Optimization, 13*, 75–93.
Relich, M. (2011a). CP-based decision support for scheduling. *Applied Computer Science, 7*, 7–17.
Relich, M. (2011b). Project prototyping with application of CP-based approach. *Management, 15*(2), 364–377.
Relich, M. (2013). Fuzzy project scheduling using constraint programming. *Applied Computer Science, 9*, 3–16.
Relich, M. (2014a). A declarative approach to new product development in the automotive industry. In *Environmental issues in automotive industry* (pp. 23–45). Berlin: Springer.
Relich, M. (2014b). A constraint programming approach for scheduling in a multi-project environment. *International Journal of Advanced Computer Science and Information Technology, 3*(2), 156–171.
Relich, M. (2015). Identifying relationships between eco-innovation and product success. In P. Golinska & A. Kawa (Eds.), *Technology management for sustainable production and logistics* (pp. 173–192). Berlin: Springer.
Relich, M. (2016). A knowledge-based system for new product portfolio selection. In P. Rozewski et al. (Eds.), *New frontiers in information and production systems modelling and analysis* (pp. 169–187). Cham: Springer.
Relich, M. (2017). Identifying project alternatives with the use of constraint programming. In *Information systems architecture and technology, advances in intelligent systems and computing* (pp. 3–13). Cham: Springer.
Sitek, P., & Wikarek, J. (2008). A declarative framework for constrained search problems. In *International Conference on Industrial, Engineering and Other Applications of Applied Intelligent Systems* (pp. 728–737). Berlin: Springer.
Soto, R., Kjellerstrand, H., Gutiérrez, J., López, A., Crawford, B., & Monfroy, E. (2012). Solving manufacturing cell design problems using constraint programming. In *Advanced research in applied artificial intelligence* (pp. 400–406). Berlin: Springer.
Srivastava, B., Kambhampati, S., & Do, M. B. (2001). Planning the project management way: Efficient planning by effective integration of causal and resource reasoning in RealPlan. *Artificial Intelligence, 131*(1-2), 73–134.
Trojet, M., H'Mida, F., & Lopez, P. (2011). Project scheduling under resource constraints: Application of the cumulative global constraint in a decision support framework. *Computers & Industrial Engineering, 61*(2), 357–363.
Tsang, E. (1996). *Foundations of constraint satisfaction*. Colchester: University of Essex.
Yang, D., & Dong, M. (2012). A constraint satisfaction approach to resolving product configuration conflicts. *Advanced Engineering Informatics, 26*, 592–602.
Yassine, A. A., Mostafa, O., & Browning, T. R. (2017). Scheduling multiple, resource-constrained, iterative, product development projects with genetic algorithms. *Computers & Industrial Engineering, 107*, 39–56.

Chapter 3
Method for Supporting Product Development

The proposed method consists of two main parts: the use of computational intelligence (CI) techniques for solving the problem stated in the forward form, and constraint programming (CP) techniques for solving the problem stated in the inverse form. Figure 3.1 illustrates these parts and their interrelations in the context of enterprise databases. As the method is dedicated to products that are modified on the basis of existing products, the data stored in enterprise databases can be used to identify relationships between input and output variables, and evaluate the product's potential.

Solving the problem stated in the inverse form requires a finite set of variables and constraints that link these variables and limit the number of solutions. Moreover, finding solutions to problems stated in the forward and inverse form is based on cause-and-effect relationships that are used for the evaluation of the product's potential and simulations related to verification of the possibility of achieving the desirable value of a project outcome.

Figure 3.2 illustrates the proposed method in the context of identifying relationships and their use to solve problems stated in the forward and inverse form. The evaluation of the potential of a new product is mainly based on economic efficiency of an NPD project.

Estimation errors are calculated for each estimation technique in training and testing sets in order to select the best predictor (with good generalisation capabilities and the least errors in the testing set). Finally, identified relationships are used to predict an output variable, e.g. the profit from a new product (the forward form of the problem), and to search possible solutions that achieve the desirable value of an output variable (the inverse form of the problem).

Enterprise databases mainly store the data in a discrete numerical form. Nevertheless, relationships between factors affecting product development are often non-linear and incoherent. The NPD process involves many factors that can indirectly impact the success of a new product. As a result, some estimation techniques are more suitable than others. For example, artificial neural networks have the

M. Relich, *Decision Support for Product Development*, Computational Intelligence Methods and Applications,
https://doi.org/10.1007/978-3-030-43897-5_3

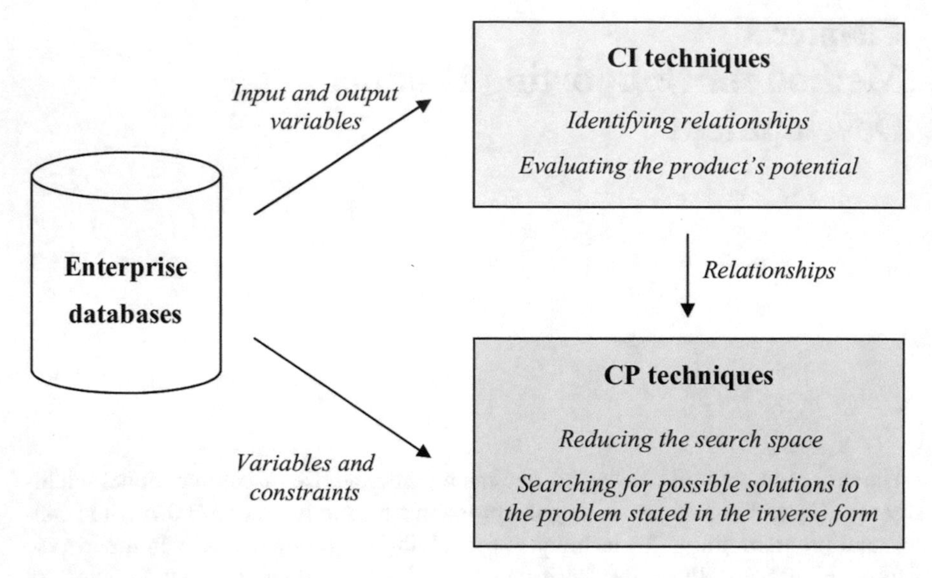

Fig. 3.1 The proposed method for solving the problem stated in the forward and inverse form

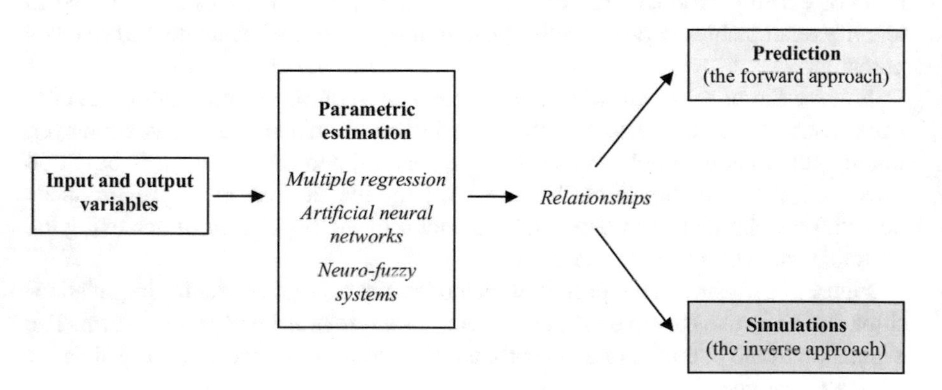

Fig. 3.2 Parametric estimation for identifying relationships

learning capability to approximate non-linear functions. In turn, neuro-fuzzy structures cope with identifying non-linear and incoherent relationships from among factors affecting product development. These techniques are related to computational intelligence that includes a set of nature-inspired computational methodologies and approaches to address complex real-world problems in which mathematical or traditional modelling can be useless (Siddique and Adeli 2013).

A Review of Selected Computational Intelligence Techniques

Computational intelligence includes a set of methodologies that are inspired by biological functions exhibited in natural systems to solve problems of information processing that are ineffective or unfeasible when solved with traditional approaches based on statistical modelling (Prieto et al. 2016). Computational intelligence is a unified and comprehensive platform of conceptual and computing endeavours of fuzzy sets, neurocomputing, and evolutionary methods (Pedrycz 2006). Since fuzzy logic, artificial neural networks, and genetic algorithms deal with numerical data, and all of them have the capability of pattern recognition and have a significant difference with respect to traditional artificial intelligence, they are the members of the computational intelligence family (Konar 2006). Compared to traditional artificial intelligence, a significant characteristic of computational intelligence is that the precise model needs not to be established when dealing with imprecise, uncertain, and incomplete information (Huang et al. 2006). Siddique and Adeli (2013) argue that computational intelligence is addressed to complex real-world problems in which mathematical or traditional modelling can be useless for the following reasons: the processes might be too complex for mathematical reasoning, it might contain some uncertainties during the process, or the process might simply be stochastic in nature. Consequently, computational intelligence is especially useful for solving problems in which valid and formalised models cannot be easily established. It is also suitable to deal with the combinational problem in designing complicated systems (Huang et al. 2006).

Computational intelligence identifies the nature of the individual technologies and exploits their highly complementary character, improving synergy between artificial neural networks, fuzzy logic, and evolutionary algorithms (Pedrycz 2006). As a result, different hybrid systems have been developed combining the advantages of the individual techniques, for example, neuro-fuzzy systems, evolutionary neural networks, and genetic fuzzy systems. These systems have been successfully used for pattern identification among datasets and prediction of future trends (Prieto et al. 2016). The next subsections present the selected intelligent technologies that can be used for evaluation of the product's potential.

Artificial Neural Networks

An artificial neural network (ANN) is a model that emulates a biological neural network. Each artificial neuron receives signals from the environment, or other artificial neurons, gathers these signals, and when fired, transmits a signal to all connected artificial neurons. The firing of an artificial neuron and the strength of the exiting signal are determined by the activation function. The artificial neuron collects all incoming signals and computes a net input signal as a function of the respective weights. The net input signal serves as input to the activation function

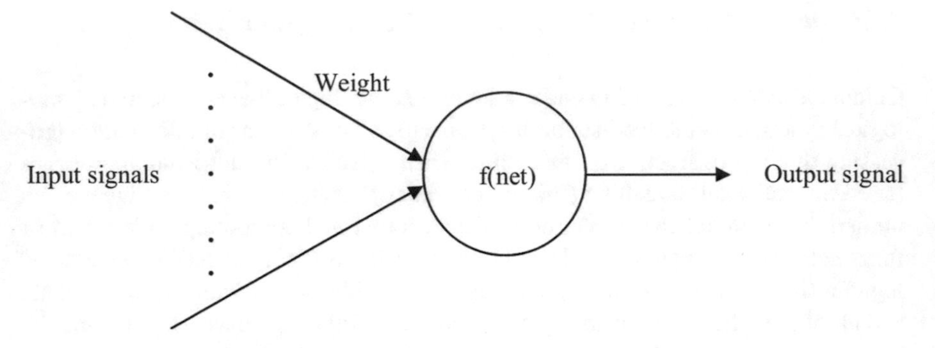

Fig. 3.3 A framework of an artificial neuron

that calculates the output signal of the artificial neuron (Engelbrecht 2007). A sigmoid activation function can squash the total input summation to a bounded output value. Figure 3.3 illustrates a framework of an artificial neuron.

An ANN is a network that usually consists of an input layer, hidden layers, and an output layer. The first layer represents input data and the last layer—the corresponding output. Between these layers, one or more intermediate (hidden) layers contain a variable number of nodes that provide sufficient complexity to the network to enable the identification of complex and non-linear relationships between inputs and outputs (Medsker 2012). The multi-layer feedforward network is the most widely studied and used neural network model in practice (Zhang 2010). An example of a multi-layer neural network is illustrated in Fig. 3.4.

Commonly, artificial neurons in one layer are fully connected to neurons in the next layer in the multi-layer feedforward network. Since a weight is associated with each connection, an ANN contains a large matrix of weights that are adjusted in the training phase. As a result, a large set of input–output pairs can be learned. Most applications use the backpropagation algorithm, or a variation of it, for training multilayer ANN (Medsker 2012). It should be pointed out that, besides multi-layer networks, several different ANN types have been developed (e.g. single-layer, temporal, self-organising, combined supervised and unsupervised ANNs), in which artificial neurons in one layer can be partially connected to neurons in the next layer and/or feedback connections to previous layers are also possible.

Neural networks are computing models for information processing and are particularly useful for identifying the complex relationships among a set of variables or patterns in the data. They have two important characteristics: parallel processing of information and learning and generalisation from experience (Zhang 2010). ANNs have been used to a wide range of applications, including data mining, image processing, pattern recognition, classification, and forecasting.

ANNs include several characteristics that make them suitable for the above-mentioned applications. First, ANNs do not require several unrealistic a priori assumptions about the underlying data generating process and specific model structures, compared to the traditional model-based methods. Moreover, ANNs are able to solve problems that have imprecise patterns or data containing incomplete and

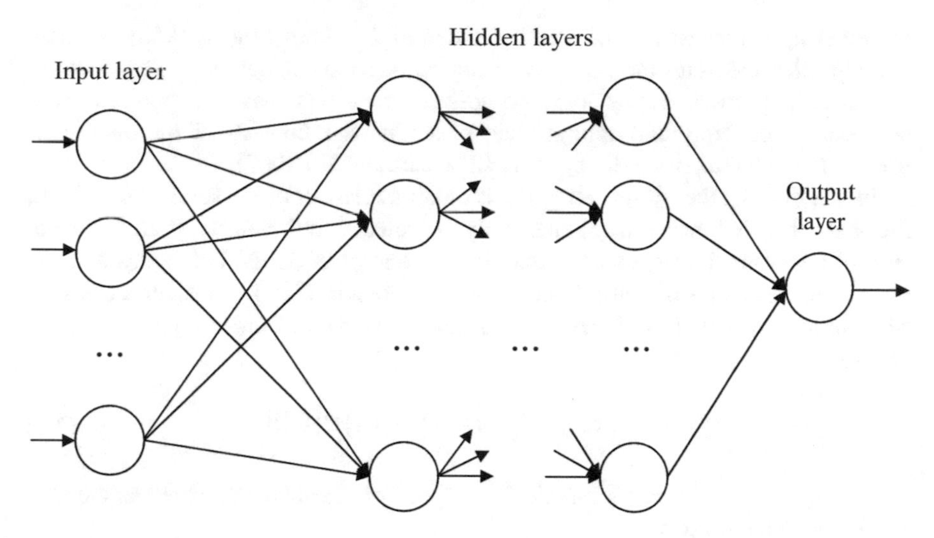

Fig. 3.4 A multilayer neural network architecture

noisy information with a large number of variables. Additionally, as real-world data and relationships are inherently non-linear, ANNs with their non-linear and non-parametric nature are more appropriate for modelling complex data mining problems than traditional linear tools (Zhang 2010).

The quality of ANN performance depends on neural network design, training algorithm, and suitable data preparation (e.g. input variable selection, data preprocessing). An architecture of neural network (e.g. the number of hidden layers, neurons in each hidden layer) and its training parameters are usually determined with the use of trial and error (generate and test) method. A well-trained network should have the generalisation ability that strongly affects quality performance of an ANN. To select the most appropriate neural network, some overall error measures (e.g. the mean squared errors—MSE or sum of squared errors—SSE) are often used to determine an objective function or performance metric. The difficulties of automatically designing and training a neural network can be considered as a potential drawback of using this computational intelligence technique.

Fuzzy Logic

The theory of fuzzy sets has emerged from limitations of traditional computing techniques that were not effective in dealing with problem description in which vagueness, imprecision, and subjectivity are immanent (Hudec 2016). Previous theories of logic had assumed that the rules of reasoning were clear and that they could be expressed in words or mathematical symbols. Fuzzy logic is much closer to human reasoning and natural language that are usually imprecise. The concept of membership values was introduced by Zadeh in the 1960s (Zadeh 1965). A membership value measures the degree or extent to which an object meets vague and/or imprecise properties. Fuzzy logic comprises fuzzy sets that can represent non-statistical uncertainty and approximate reasoning that consists of the operations used to make inferences in fuzzy logic (Eberhart and Shi 2007).

In classical set theory, an element x is either a member of the set A or not. Thus, the characteristic function $\mu_A(x)$ takes only the values 1 or 0. Fuzzy sets are an extension of two-valued (crisp) sets to handle the concept of partial belonging that can express uncertainties of natural language (Engelbrecht 2007). Consider a classical set X of the universe U. A fuzzy set A is defined as a set of ordered pairs, a binary relation:

$$A = \{(x, \mu_A(x))|, x \in X|, \mu_A(x) \in [0,1]\} \tag{3.1}$$

where $\mu_A(x)$ is a membership function that specifies the degree to which an element x belongs to the fuzzy set A.

Classical sets can be considered as a special case of fuzzy sets with all membership degrees equal to 1. The shapes of membership functions are determined by experts corresponding to their subjective judgement in the respective domain. Figure 3.5 illustrates shapes of commonly used membership functions in most practical applications, such as triangular (a), trapezoidal (b), Gaussian (c), and sigmoid (d).

The operations on fuzzy sets are defined through operations on their respective membership functions. Basic operations on fuzzy relations include equality,

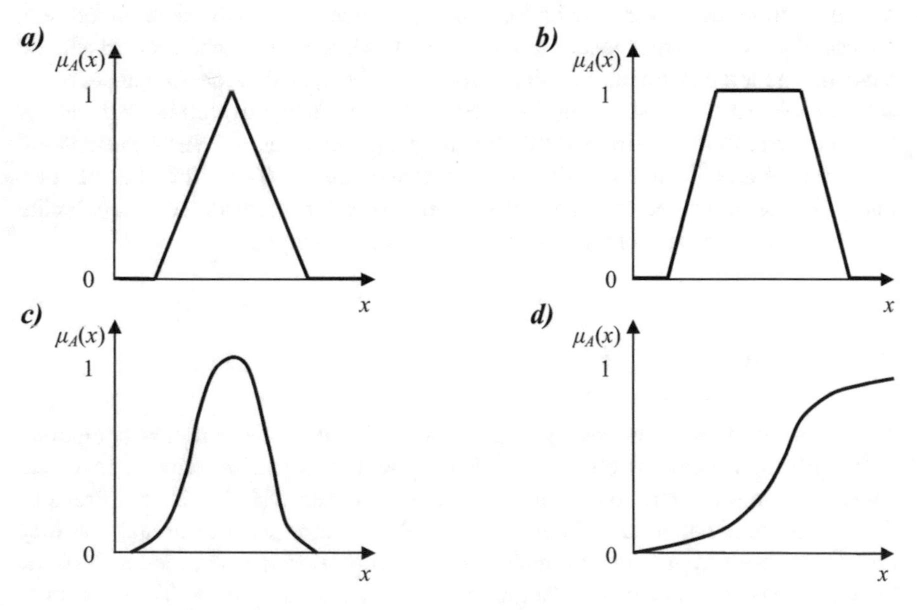

Fig. 3.5 Shapes of commonly used membership functions

inclusion, complement, intersection, and union (Engelbrecht 2007; Hudec 2016). Fuzzy logic can be considered as an extension of infinite-value logic from the perspective of incorporating fuzzy sets and fuzzy relations into the system of infinitive-valued logic. Fuzzy logic is related to linguistic variables in natural language and aims to provide foundations for approximate reasoning with imprecise propositions (Bojadziev and Bojadziev 2007). Linguistic variables can be used in financial and management systems to describe variables such as risk, investment, stress, confidence, truth, income, profit, cost, or project duration (Carlsson et al. 2004; Gil-Lafuente 2005; Bojadziev and Bojadziev 2007; Relich 2012; Awasthi et al. 2014). Figure 3.6 illustrates an example of linguistic values (labels) that are related to the linguistic variable of product attractiveness.

The dynamic behaviour of a system can be described with the use of a set of linguistic fuzzy rules. These rules are based on the knowledge and experience of a domain expert, or they are determined during adjustment of the system parameters. A fuzzy rule consists of the antecedent(s) and consequent(s) that are propositions containing linguistic variables:

$$\textbf{If } x_1 \text{ is } A_1 \text{ and} \ldots \text{and } x_n \text{ is } A_n \textbf{ then } y_1 \text{ is } B_1 \text{ and} \ldots \text{and } y_m \text{ is } B_m \tag{3.2}$$

where A_1, ..., A_n and B_1, ..., B_m are linguistic values of x_1, ..., x_n and y_1, ..., y_m, respectively. The antecedents of a rule combine fuzzy sets through the use of the logic operations (e.g. complement, intersection, union) (Engelbrecht 2007). Two of the most common fuzzy if-then rules refer to the Mamdani-type fuzzy rule (the consequent part of a rule is a fuzzy variable) and Takagi–Sugeno–Kang model (the consequent is a polynomial function of the inputs) (Eberhart and Shi 2007). Figure 3.7 presents a framework of a fuzzy rule-based reasoning system.

Fuzzification aims to find a fuzzy representation of crisp input values. This is achieved through the use of the membership functions associated with each fuzzy

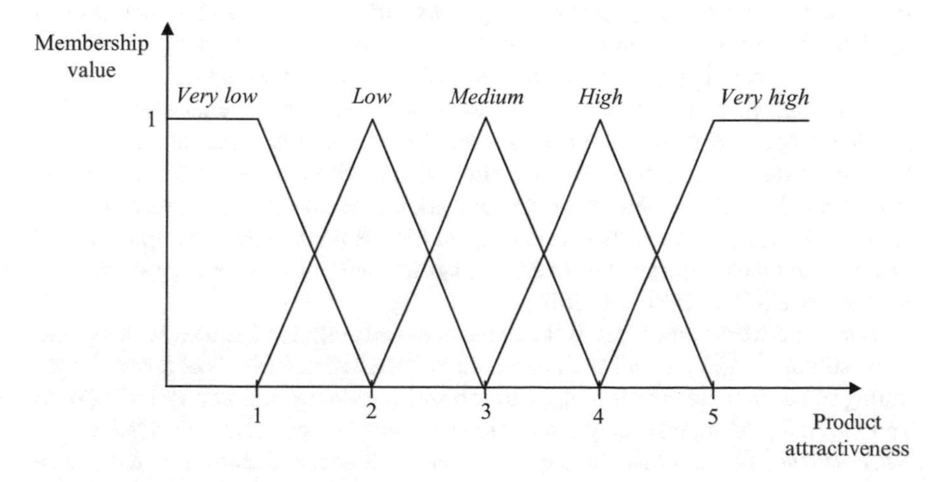

Fig. 3.6 An example of linguistic values

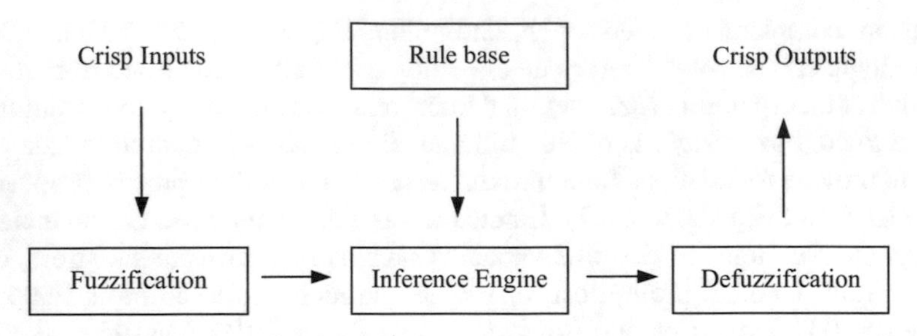

Fig. 3.7 Fuzzy rule-based reasoning system

set in the rule input space. The inferencing process is concerned with mapping the fuzzified inputs (received from the fuzzification process) to the rule base and producing a fuzzified output for each rule. In turn, defuzzification aims to convert the output of the fuzzy rules into a crisp value (Engelbrecht 2007). The several defuzzification methods take into account the shape of the clipped fuzzy numbers (e.g. length of supporting intervals, height of the clipped triangles and trapezoids, closeness to central triangular numbers) and the computational complexity (Bojadziev and Bojadziev 2007). A crisp output value can be calculated according to maximum value strategies that consider only flat segment or point with the maximal value of height, and gravity strategies that consider the entire shape of the membership function (e.g. the centre of gravity method) (Hudec 2016).

Neuro-fuzzy Systems

Neuro-fuzzy systems combine methods from neural network theory with fuzzy systems. The term *neuro-fuzzy systems* is often used interchangeably with terms *neural-fuzzy*, *fuzzy-neural*, or *fuzzy-neuro systems*. Mitra distinguishes neuro-fuzzy hybridisation as *fuzzy-neural network* that is a neural network equipped with the capability of handling fuzzy information and *neural-fuzzy system* that is a fuzzy system augmented by neural networks to enhance some of its characteristics like flexibility, speed, and adaptability (Mitra and Hayashi 2000). An example of neural-fuzzy hybridisation is a set of fuzzy neurons that realise the common operations of fuzzy set theory (union, intersection, aggregation). Neuro-fuzzy systems can endow fuzzy inference engines with learning capabilities using the neural components, for example, to tune the parameters of a fuzzy system and/or to compile the structure of the rule base (Castellano et al. 2007).

Neural networks and fuzzy systems have several common features such as function estimators and parallel implementations (Medsker 2012). Feedforward computing in neural networks is similar to forward reasoning in fuzzy systems (Wang and Fu 2006). Moreover, they are numerical, trainable, and dynamic systems that can estimate a function without any mathematical model and learn from experience with sample data. A fuzzy system adaptively infers and modifies its fuzzy associations from representative numerical samples, and enables explicit knowledge representation (Mitra and Hayashi 2000). In turn, neural networks have adaptive, optimising, fault tolerance, and generalising properties, and they are able to identify complex non-linearity relationships among data.

The identified relationships are reflected in the structure of a neural network (in its weights and biases). The difficulty of interpreting the output of a trained network is the reason why users call them sometimes "black boxes" (Wang and Fu 2006). The extraction of rules from neural networks can facilitate users to understand the estimation process, because rules are a form of knowledge that humans can easily verify, transmit, and expand. Fuzzy systems provide a powerful framework for expert knowledge representation, whereas neural networks provide learning capabilities and exceptional suitability for computationally efficient implementation (Mitra and Hayashi 2000). As a result, the integration of neural and fuzzy systems leads to a symbiotic relationship that enables the generation of fuzzy rules from training data.

Neuro-fuzzy systems can use linguistic information from the domain expert and/or measurement data. If these systems use only measurement data, then the method of identification is referred to as automatic rule generation (Czogala and Leski 2000). In general, the learning ability of a neuro-fuzzy system should include a rule generation algorithm and a method of parameter tuning (Rutkowska 2002).

One of the most popular neuro-fuzzy systems is an adaptive neuro-fuzzy inference system (ANFIS). Fuzzy sets are modelled by bell-shaped membership functions that can be easily used for learning a neural network. ANFIS uses backpropagation to learn the antecedent parameters (membership functions) and least mean square estimation to determine the coefficients of the linear combinations in the conclusions of rules. After learning a neuro-fuzzy network, the obtained model is usually interpreted as a system of fuzzy rules (Nauck and Nürnberger 2013). ANFIS encodes fuzzy if-then rules of Takagi–Sugeno type in the following form:

$$R_r : \textbf{If}\ x_1 \text{ is } A_1 \text{ and} \ldots \text{and } x_n \text{ is } A_n\ \textbf{then}\ y = \alpha_0^r + \alpha_1^r x_1 + \cdots + \alpha_n^r x_n \quad (3.3)$$

Neuro-fuzzy systems have been successfully employed in many fields of business, for example, manufacturing (Seyedhoseini et al. 2010; Fazlollahtabar and Mahdavi-Amiri 2013), financial prediction (Relich 2008; Wang et al. 2011), supply chain (Gumus et al. 2009; Latif et al. 2014), customer satisfaction (Kwong et al. 2009), and decision-making (Efendigil et al. 2009; Relich 2010). The review of research articles dedicated to the use of neuro-fuzzy systems in business indicates that most applications are concerned with manufacturing and distribution problems, whereas a few articles refer to business planning, including strategic and scenario planning (Kar et al. 2014; Relich 2016; Relich and Pawlewski 2017; Rajab and Sharma 2018).

Computational Intelligence for Evaluating NPD Projects

Solving the problem stated in the forward form refers to evaluation of the potential of a new product using identified relationships between input and output variables. These relationships are also used in solving the problem stated in the inverse form, namely in seeking possibilities to fulfil NPD project requirements. As an inverse approach aims to search possible alternatives for reaching the desirable value of an output variable, it involves some indirect variables that are used for estimation of decision variables for evaluating the potential of a new product. Hence, an inverse approach is related to parametric estimation instead of analogical reasoning. A framework of the parametric estimation method for evaluating the potential of a new product is presented in Fig. 3.8.

An estimation model is determined according to the least error obtained on the testing dataset by multiple regression or artificial neural networks. Input variables refer to product attributes (e.g. shape, size, weight, power consuming), enterprise (e.g. the NPD project budget and deadline, production volume, marketing budget), and its environment (e.g. the cost of materials, sales volume). An output variable can be related to the cost of an NPD project, profit from a new product, its life cycle, etc. Taking into account that among the data there are also ongoing products, for which life cycle and return on investment are hard to evaluate, the profit from a new product in the first year after launch is assigned as the output variable.

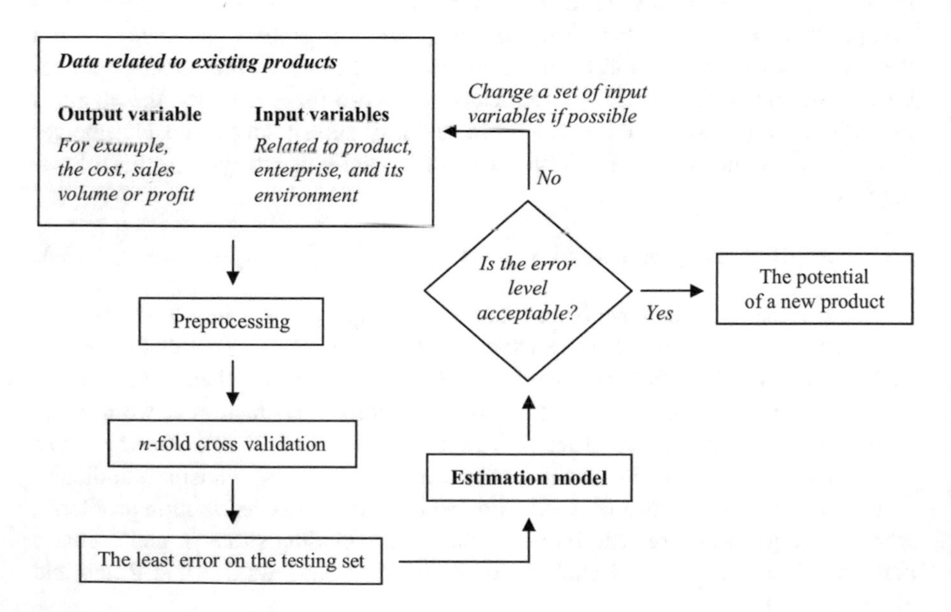

Fig. 3.8 Parametric estimation for evaluating the potential of a new product

Parametric Estimation Methods

This study considers NPD projects that are modifications of existing products. As a result, estimation methods for evaluating the potential of a new product can be based on the data from an enterprise database referring to previous similar NPD projects. This evaluation can be carried out through analogical reasoning and parametric estimation methods. Analogical reasoning uses the actual performance of a previous similar project to assess the duration, cost, or profit of a new project. Parametric estimating uses an algorithm (including variables needed) to identify statistical relationships between historical data from previous similar projects, and use these relationships to estimate performance of an NPD project. The identified cause-and-effect relationships can then be used to simulations towards searching for possible solutions of the problem stated in an inverse form. Consequently, parametric estimating for evaluating the potential of a new product is further considered.

In the existing literature, parametric estimation mainly occurs in the context of using multiple regression and machine learning algorithms, including neural networks to evaluate the product's potential. Song and Parry (1997) present the use of ordinary least squares regression to estimate profitability, sales, and market shares. In their estimation model, they consider factors referring to the NPD process (e.g. technical development proficiency, product tests, and commercialisation proficiency), company (e.g. marketing skills, internal commitment, cross-functional integration), and its environment (e.g. market potential, competitive intensity). Narver et al. (2004) propose a hierarchical regression model for evaluating new product success using factors regarding organisation climate and innovation orientation. In turn, Lee et al. (2014) propose an approach to the pre-launch forecasting of new product demand based on multivariate linear regression and machine learning algorithms. They build an estimation model on the basis of multiple product attribute variables classified in dimensions such as industry (e.g. the availability of complementary goods), market (e.g. the price level, number of potential customers), technology (e.g. the degree of newness, easiness of imitation, degree of functional variety), and use (e.g. the necessity of learning, frequency and duration of usage).

Many parametric estimation methods use various types of artificial neural networks (ANN) to evaluate the product's potential. Bode et al. (1995) present the results of the application of backpropagation three-layer perceptron to cost estimation problems in product design using input data such as product characteristics (e.g. size), improvement target, technical difficulty, and production volume. Li (2000) proposes a hybrid intelligent system that consists of ANN and fuzzification component, and aims to support decision-makers in marketing strategy. This system uses input variables referring to market attractiveness and business strengths factors. Seo et al. (2002) describe approximate estimation of the product life cycle cost using ANN in conceptual design. Their model consists of life cycle cost factors (e.g. market recognition, development, materials, energy, facilities, wages, packaging, transportation, storage, breakage, warranty service) and a set of product attributes (e.g. durability, functionality, volume, selling price, use time, recyclability, reusability). In turn, Huang et al. (2006) use a neural network for evaluating the value of an economic indicator on the basis of the following input variables: design cost, long life, simple structure, and convenient for manufacture and maintenance.

The comparison of multiple regression with ANN in the context of cost estimation of an NPD project has been presented by Relich (2016). The cost of product development is estimated through factors regarding product design, prototype manufacturing, and prototype tests. Product design includes factors such as the number of client's requirements translated into product specification, ideas elaborated for a new product, and components in a new product. Prototype manufacturing is described by the following factors: the number of components for assembling/processing, assembly/processing time, assembly machines, and time for assembly machine configuration. In turn, prototype tests contain factors referring to the number of components in a new product, cycles and devices for testing, and the number of suppliers for the required materials and components. Moreover, Relich (2016) proposes an approach that involves the cost of desirable reliability in the cost estimation of a new product. This approach uses parametric models for estimating the cost of an NPD project, production, and product warranty (in relation to product reliability and the relevant cost of prototype tests). The parametric cost estimation and identified relationships aim evaluation of the product's potential and, finally, the selection of an optimal portfolio of NPD projects.

Multiple regression analysis (MRA) enables the prediction of the change of dependent variable(s) taking into account given values of independent variables. The general multiple regression model is as follows:

$$Y = B_0 + B_1 X_1 + \cdots + B_i X_i + \cdots + B_n X_n \quad (3.4)$$

where:

Y dependent variable
X_i independent variables
B_0 constant
B_i unstandardised coefficients

In the presented regression model, the change of an independent variable at 1 unit results in the change of the dependent variable at the unstandardised coefficient of the changed independent variable.

The number of inputs of a neural network is the same as for the variables of the regression model, and the structure of a neural network consists of three layers. Attributes related to a new product are led to the first layer, whereas the last layer corresponds to NPD project performance. A feedforward neural network trained according to the backpropagation algorithm usually uses sigmoid function to calculate the output of each neuron except for input neurons (Kim et al. 2004). The proposed feedforward network consists of one hidden layer of non-linear sigmoid neurons followed by an output layer of a linear neuron. The hidden layer and output layer have logarithmic sigmoid and linear transfer function, respectively.

Many researchers have proven the usefulness of neural network models to cost estimation (e.g. Finnie et al. 1997; Cavalieri et al. 2004; Kim et al. 2004; Paliwal and Kumar 2009; Duran et al. 2012; Relich and Pawlewski 2018). The authors have highlighted the experimental nature of their research. Neural network theory provides only a general methodology for designing network structure, and setting parameters related to neural network training such as learning rate and momentum. Determining the number of hidden layers, neurons in each hidden layer, and parameter values in a learning algorithm is commonly the trial and error approach that aims to select an optimal setting (Tkac and Verner 2016). Consequently, the performance of ANNs is compared with other estimation techniques in the context of their credibility and accuracy (Smith and Mason 1997; Zemouri et al. 2010; Duran et al. 2012; Relich and Bzdyra 2015).

Data preprocessing is concerned with preparing the final training set that should not include irrelevant and redundant information or unreliable data. The use of data transformation and normalisation facilitates artificial neural networks to identify patterns among data. Transformation involves changing raw data inputs to create input to a network, while normalisation is a transformation performed on a single data input to distribute the data evenly and scale it into an acceptable range for the neural network. The goal of normalising data is to ensure the roughly uniform statistical distribution of values for each input and output neurons. Moreover, the values should be scaled to match the range of the input neurons, e.g. from 0 to 1 or from −1 to 1. The impact of data preprocessing and dimension reduction in the context of NPD has been verified in (Relich and Muszyński 2014).

The effectiveness of the trained ANN is verified using a dataset that was not used in the training phase. Therefore, the data are divided into the training and testing set. The testing phase helps prevent over-training of the network that can lead to the overfitting problem. The main approach used for this purpose is the cross validation (Suthaharan 2016). In the n-fold cross validation, the entire dataset is divided into n disjointed subsets of equal sizes. Furthermore, $n - 1$ folds are used for training the model, and one fold is selected for testing. For example, in fivefold cross validation, the model is trained with the use of the first four folds and tested for the fifth fold. In the second step, the fourth fold is selected for testing, and folds from 1 to 3 and 5 are used for training the model. The procedure is repeated five times as illustrated in Fig. 3.9. The n-fold cross validation provides n errors between the estimated and actual value for each training and testing set. Finally, an error related to an estimation model is calculated as the average for n-fold errors.

The performance of MRA and ANN is evaluated through using the mean absolute percentage error (MAPE) in the following form:

$$\mathrm{MAPE} = \frac{\sum \left| \frac{C_\mathrm{A} - C_\mathrm{E}}{C_\mathrm{A}} \right| \times 100\%}{n} \quad (3.5)$$

where:

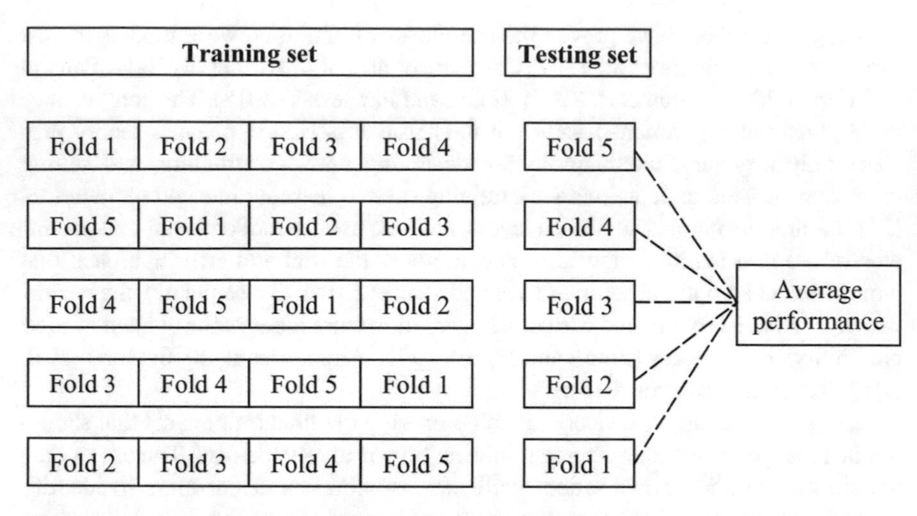

Fig. 3.9 The procedure of fivefold cross validation

C_A the actual value of NPD project performance
C_E the estimated value of NPD project performance
n the number of cases in the training/testing set

The MAPE is compared among estimation models, and then results for a model with the least MAPE in the testing set are presented to the decision-maker. Consequently, the most suitable model for a given dataset is always selected.

An Example of Evaluating the Potential of a New Product

The database includes 37 cases that are related to product development before and after the launch. Before the launch, new product performance is measured by costs related to specific phases of the NPD process such as market analysis, concept generation, product design, and prototype tests. Product performance after the launch can be referred to as sales revenue and costs of production and promotion. There is the following set of output variables: the cost of market analysis (Cma), concept generation (Ccg), product design (Cpd), prototype tests (Cpt), marketing campaign of a new product (Cmc), unit production cost (Upc), and sales volume (Sv). The set of input variables ($X_1, \ldots, X_{22}$) is presented in Chap. 2. The following relationships between input and output variables are sought:

$\text{Cma} = f(X_1, X_2, X_3, X_4)$
$\text{Ccg} = f(X_5, X_6, X_7, X_8)$
$\text{Cpd} = f(X_8, X_9, X_{10}, X_{11}, X_{12}, X_{13})$
$\text{Cpt} = f(X_{13}, X_{14}, X_{15}, X_{16}, X_{17}, X_{18})$
$\text{Cmc} = f(X_{19}, X_{20})$
$\text{Upc} = f(X_{21}, X_{22})$
$\text{Sv} = f(X_1, \ldots, X_{22})$

As past and ongoing products can have different life cycles, sales volume and relevant net profit (Np) from a new product are measured in the first year after the launch. The product price (Up) is limited by substitutive products. The potential of a new product is calculated according to Eq. (2.16).

The estimation process is carried out for each output variable separately using MRA and ANNs. The case base was randomly divided into the training set (30 cases) and testing set (7 cases). The tests were performed with the use of fivefold cross validation. Weight and bias values of an ANN have been updated with the use of the gradient descent with an adaptive learning rate algorithm (ANN-GDA) and Levenberg–Marquardt algorithm (ANN-LM). The optimal number of hidden neurons has been selected according to the trial and error approach, taking into account the minimal MAPE. The MAPE is calculated as the average of 20 simulations for each structure of a neural network with a number to the extent of 20 hidden neurons. Table 3.1 presents the MAPE in the training and testing set for fivefold cross validation ($F1, \ldots, F5$) in the context of different estimation models for the cost of product design (Cpd). An ANN trained according to the LM algorithm has achieved the least error in the training and testing data.

Figure 3.10 illustrates the distribution of the MAPE for 35 cases (7 cases in the testing set for fivefold cross validation). Differences between the presented models were verified with the use of a *t*-test at the 0.05 significance level. Table 3.2 shows the *p*-values between the presented estimation models. The results indicate the rejection of the null hypothesis that verifies whether two data vectors are from population with equal means at the given significance level.

The presented *p*-values indicate that ANNLM significantly differs from the MRA model for the testing data. As ANNLM generated the least error in the testing data, it is used to estimate the cost of product design.

There are the following values describing product design: the rate of similarity between existing products and a new product—80%, the rate of customers' needs translated into technical specification—70%, the number of parts in a new product—55, the number of employees involved in product design—4, the duration of product design—4 months, and the number of adjustments in the product specification to reduce the cost and/or time of manufacturing a new product—3. After inputting these values to the trained network, the cost of product design is predicted at 168.4 thousand €.

Table 3.1 The MAPE for the cost of product design

Model		*F*1	*F*2	*F*3	*F*4	*F*5	Average
Training set	MRA	2.26	2.30	2.71	2.66	2.58	2.50
	ANNGDA	2.82	2.74	3.94	3.84	3.65	3.40
	ANNLM	1.65	1.91	2.31	2.15	2.39	2.08
Testing set	MRA	3.86	3.69	2.62	2.28	3.32	3.15
	ANNGDA	2.74	1.67	0.91	2.90	3.75	2.39
	ANNLM	3.63	1.63	0.34	1.92	2.71	2.05

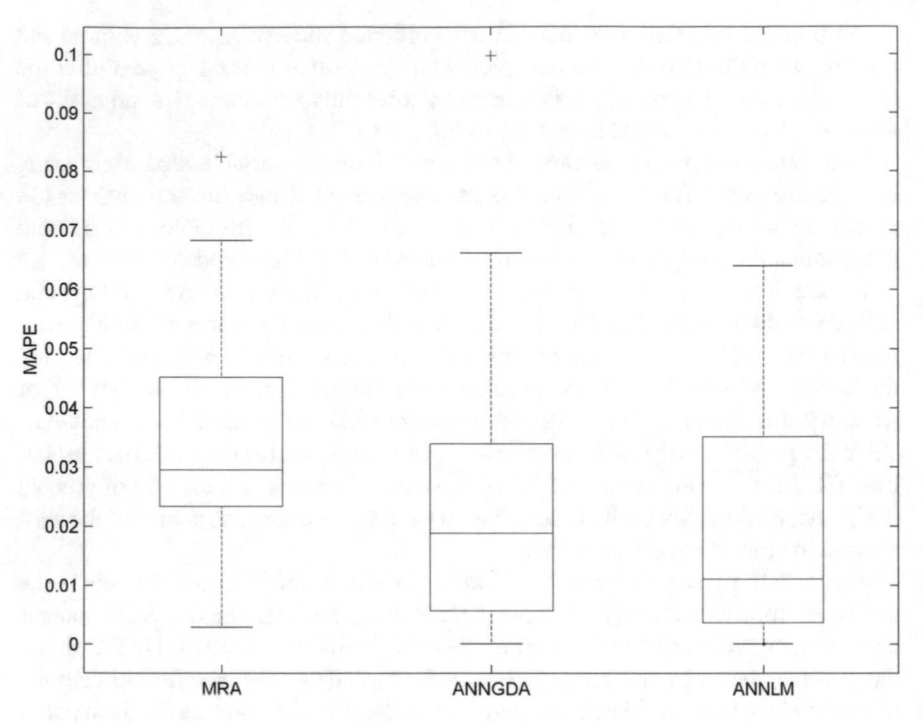

Fig. 3.10 Box plot for estimation results

Table 3.2 The p-values between different estimation models

Model	MRA	ANNGDA	ANNLM
MRA	1		
ANNGDA	0.1701	1	
ANNLM	0.0314	0.5359	1

The cost related to other product development phases is as follows: market analysis—41.9 thousand €, concept generation—23.7 thousand €, prototype tests—90.1 thousand €, marketing campaign—172 thousand €, unit production cost—32.73 €, unit price of a new product—69.99 €, and sales volume in the first year after the launch—25 thousand units. The expected profit in the assumed time period reaches 435.4 thousand €.

An Example of Evaluating the Customers' Satisfaction

The customers' satisfaction is an example of variable that can be successfully described in the imprecise form, using the fuzzy set theory. The customers' opinions can be interviewed individually through the questionnaire. The customers' satisfaction contains the aspect of four areas of the marketing mix (product, price, place, and promotion).

The customers' satisfaction is measured with the use of a five-point response anchor numbered from 1 to 5 (from very low = 1 to very high = 5). It is assumed that respondents have suitable knowledge about a product and have used once previously. The variables related to product are as follows: shape (very usual = 1 to very modern = 5), degree of power consumption (very high = 1 to very low = 5), and easiness of use (very low = 1 to very high = 5). The price is measured according to the following scale: below 100 € = 1, from 100 to 200 € = 2, from 200 to 300 € = 3, from 300 to 400 € = 4, and above 400 € = 5. The variable related to place is measured through the social class (income) of the respondent as follows: below 500 € = 1, from 500 to 1000 € = 2, from 1000 to 1500 € = 3, from 1500 to 2000 € = 4, and above 2000 € = 5. The variable related to promotion refers to the number of monthly instalments as follows: below 4 months = 1, from 5 to 6 months = 2, from 7 to 8 months = 3, from 9 to 10 months = 4, and from 11 to 12 months = 5.

The structure of an ANFIS implemented in the MATLAB® software is illustrated in Fig. 3.11. First, the ANFIS generates a Sugeno-type fuzzy inference structure based on the given number and types of membership functions. In the analysis, the subtractive clustering is applied to extract rules. The subtractive clustering algorithm is a fast one-pass and robust method for estimating the number and centre of clusters in a dataset. This algorithm is particularly useful if the user is not familiar with the number of clusters that should be distinguished among data.

Table 3.3 presents the comparison of the ANFIS with commonly used regression analysis. The results indicate that the ANFIS model outperforms regression analysis in both the training and testing datasets.

The proposed approach is based on estimation models related to multiple regression and neuro-fuzzy systems. Estimation models are compared with the use of the

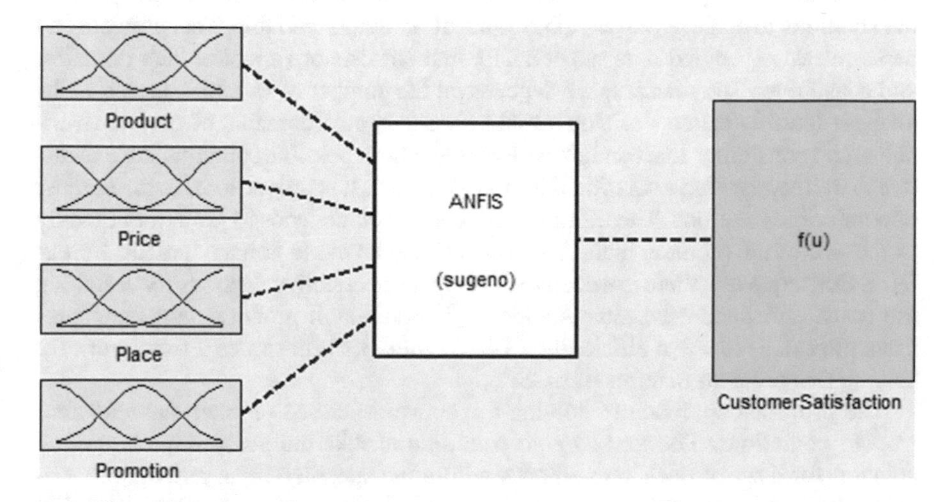

Fig. 3.11 Structure of a neuro-fuzzy system

Table 3.3 The MAPE for the customers' satisfaction

Model		*F1*	*F2*	*F3*	*F4*	*F5*	Average
Training set	MRA	2.41	2.31	2.38	2.49	2.48	2.41
	ANFIS	1.83	1.64	1.77	2.15	1.97	1.87
Testing set	MRA	8.08	6.17	7.40	10.21	9.38	8.25
	ANFIS	6.53	5.09	5.74	7.61	7.83	6.56

average error between the estimated and actual data within n-fold cross validation. Finally, the best model (with the least error) is selected to estimate the performance of a new product project.

The presented method for evaluating the product's potential begins with preparing available data related to a new product. The data can be acquired from an enterprise system (e.g. ERP, CRM, CAD software) and the market (e.g. customer requirements for a new product). The sales and marketing department analyses customer responses about existing products, and the R&D department creates concepts for a new product taking into account customers' opinions. An estimation model is built with the use of ANN and ANFIS and compared to MRA in order to verify its accuracy and applicability in the problem. Finally, the estimation model is applied to predict the value of an output variable. Moreover, cause-and-effect relationships are used for simulations that support the decision-maker in solving NPD-related problems stated in the inverse form.

Constraint Programming for Simulating NPD Project Performance

The solution of the problem stated in the inverse form depends on searching possible solutions to achieve the desirable value of an output variable. This problem can be seamlessly specified in terms of a CSP that consists of variables, their domains, and constraints. The search space depends on the number of variables chosen to the analysis (further called decision variables), a range of domains of decision variables, and constraints that can link variables and limit possible solutions. An exhaustive search always find a solution if it exists, but its cost is proportional to the number of admissible solutions. Therefore, an exhaustive search tends to grow very quickly as the size of the problem increases, which limits its usage in many practical problems. Consequently, there is a need to develop more effective methods for searching the space and finding possible solutions. This research proposes constraint programming (CP) to solve efficiently a CSP. Figure 3.12 illustrates a framework for solving this problem in terms of a CSP.

The proposed approach of solving the above-described problem includes four stop/go conditions. The first stop/go condition checks the possibility of existing solution for a set of decision variables within the specified input and output variables, their domains, and constraints (1). If there is no solution, then the possibility

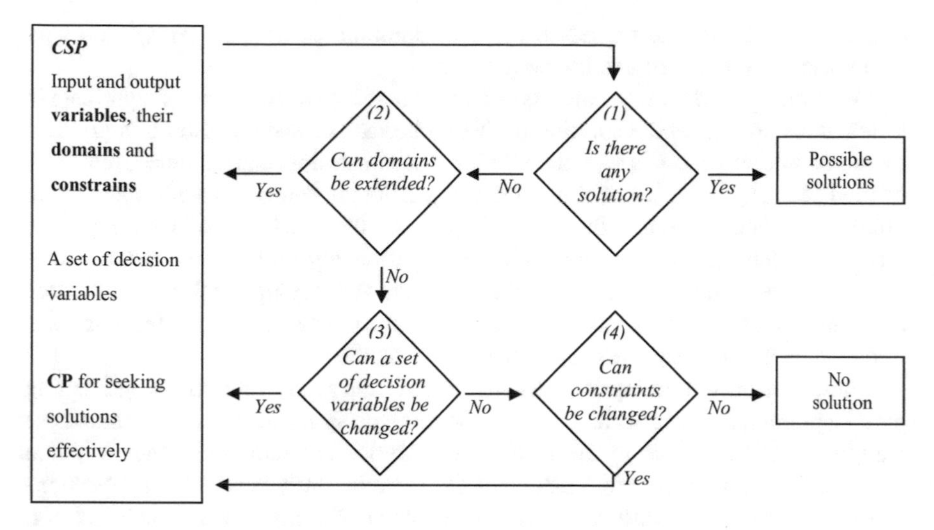

Fig. 3.12 A framework for solving the problem stated in the inverse form

to extend domains related to the selected decision variables is verified (2). If domains of decision variables cannot be extended, then the possibility to change the set of decision variables is verified (3). If this set cannot also be changed, then the possibility to change constraints is verified (4). Finally, if any constraints cannot be changed, then the last stop condition leads to the empty solution set.

Constraint Programming Techniques

Constraint programming is an alternative approach to programming that embraces reasoning and computing (Apt 2003). The main notion of this concept is a constraint on a sequence of variables that is related to their domains. In other words, it is a requirement that determines specific combinations of values from the variable domains. A given problem should be formulated as a constraint satisfaction problem to solve it with the use of CP. This can be performed through introduction of some variables ranging over specific domains and constraints over these variables, and selection of some language in which constraints are expressed.

In general, more than one specification of a problem as a CSP exists. The chosen specification can be solved with the use of domain-specific methods, general methods, or a combination of both. The domain-specific methods are usually provided as implementations of special-purpose algorithms (e.g. in the form of a program for solving systems of linear equations, an implementation of the unification algorithm). In turn, the general methods are concerned with reducing the search space and with specific search methods (Apt 2003). The algorithms dedicated to reduce the search space are usually called constraint propagation algorithms. They achieve various forms of local consistency towards approximating the notion of global consistency. The top-down search methods combine constraint propagation with backtracking and branch and bound algorithms.

Constraint programming consists of two phases: the specification of a problem in terms of constraints and a solution of this problem. The specification of a problem by means of constraints is very flexible because constraints can be added, removed, or modified (Apt 2003). CP aims to develop efficient domain-specific methods in order to use them instead of the general methods. They can be used to develop more efficient constraint solvers, constraint propagation algorithms, and search algorithms. Consequently, CP is a powerful paradigm for solving combinatorial search problems, in which the user declaratively states the constraints on the feasible solutions for a set of decision variables (Rossi et al. 2006).

Constraint programming is qualitatively different from the other programming paradigms such as object-oriented and concurrent programming. Compared to these paradigms, CP is closer to the ideal of declarative programming (Van Roy and Haridi 2004). The declarative nature of CP is particularly useful for applications where it is enough to state *what* has to be solved without saying *how* to solve it (Banaszak et al. 2009). For instance, it verifies whether an enterprise has the capability (available resources, experience, etc.) to satisfy constraints related to project requirements (completion time, cost, etc.). In the case of a positive answer, the most efficient way to execute a project is sought. The idea of constraint-based programming is to solve problems through simply stating constraints that must be satisfied by a solution of the problem (Frühwirth and Abdennadher 2003).

The advantages of using CP refer to declarative problem modelling, propagation of the effects of decisions by means of efficient algorithms, and search of optimal solutions. Specific search methods and constraint propagation algorithms used in CP enable significant reduction of the search space. Consequently, CP is suitable for modelling complex problems, e.g. related to resource allocation.

An Example of Searching for Desirable Project Performance

An example consists of four variants: (1) a basic variant for originally declared decision variables, their domains, and constraints; (2) an extension of domains for the selected decision variables; (3) a change in the set of decision variables; and (4) a change in the set of constraints.

A Basic Variant

Let us assume that the expected profit from a new product equals 435.4 thousand € (see an example of evaluating the potential of a new product), and it does not fulfil the decision-maker's expectations. The decision-maker is interested in increasing the profit to 440 thousand €. To verify the possibility of existing solutions, this problem is stated in the inverse form. The solution of the problem stated in the inverse form uses constraint programming, and requires the specification of decision variables, their domains, and constraints, among which are cause-and-effect relationships between variables (e.g. their mutual impact on the cost or sales).

There are the same variables (X_1, …, X_{22}) and constraints as in the forward approach (see an example of evaluating the potential of a new product). An estimation method identifies the relationships between input and output variables, including the increase in the total cost (TC) and sales volume (Sv) by a unit change in the input variable (see Table 3.4). The total cost of launching a new product is calculated according to Eq. (2.15).

The increases presented in Table 3.4 can facilitate the selection of decision variables that are used within the inverse approach. The following variables have been selected as decision variables: the rate of customers' needs translated into technical specification (X_9), the number of employees involved in product design (X_{11}), the duration of elaborating product design (X_{12}), the number of prototype tests (X_{15}), the number of employees involved in prototype tests (X_{17}), and the duration of prototype tests (X_{18}). Domains for these decision variables are as follows: X_9 = 70%#80%, X_{11} = 3#5, X_{12} = 4#6, X_{15} = 8#10, X_{17} = 2#3, and X_{18} = 3#5.

In addition to constraints regarding relationships between input and output variables (Cma, …, Sv), there are also constraints related to:

- The number of R&D team members

$$X_{11} + X_{17} \leq 6$$

- The duration of product design and prototype tests (in months)

$$X_{12} + X_{18} \leq 7$$

- The rate of customers' needs translated into technical specification

$$X_9 \geq 70\%$$

- The number of prototype tests

$$X_{15} \geq 8$$

Table 3.4 Increments in the cost and sales volume

	ΔTC	ΔSv	ΔSv/ΔTC		ΔTC	ΔSv	ΔSv/ΔTC
X_1	0.00028	0.0004	1.43	X_{12}	0.00067	0.0039	5.82
X_2	0.00089	0.0012	1.35	X_{13}	0.00073	0.0024	3.29
X_3	0.00040	0.0007	1.75	X_{14}	0.00024	0.0009	3.75
X_4	0.00124	0.0041	3.31	X_{15}	0.00002	0.0001	4.95
X_5	0.00021	0.0030	1.43	X_{16}	0.00001	0.00004	4.09
X_6	0.00001	0.00004	4.10	X_{17}	0.00086	0.0054	6.28
X_7	0.00148	0.0067	4.53	X_{18}	0.00063	0.0041	6.51
X_8	−0.00084	−0.0011	1.31	X_{19}	0.00176	0.0063	3.58
X_9	0.00027	0.0013	4.81	X_{20}	0.00001	0.00003	3.25
X_{10}	0.00017	0.0004	2.35	X_{21}	0.00086	0.0041	4.77
X_{11}	0.00089	0.0057	6.40	X_{22}	0.00048	0.0022	4.58

Prototype tests should ensure appropriate reliability of a new product to decrease the potential warranty cost and, finally, increase customer satisfaction. Product reliability can be estimated through measures such as the number of usage cycles for the first failure. Interrelations between the potential warranty cost and the number of usage cycles for the first failure (the cost of prototype tests), as well as its involvement in the total cost of investment, are presented in Relich (2016).

The last constraint refers to the desirable profit from a new product (in thousand €):

$$Np \geq 440$$

The problem stated in the inverse form has been implemented in Mozart/Oz, i.e. a multiparadigm programming language released with an open-source licence. Mozart/Oz contains most of the major programming paradigms, including logic, functional, imperative, object-oriented, concurrent, constraint, and distributed programming. The significant strengths of Mozart/Oz are constraint and distributed programming that are able to effectively solve many practical problems, for example reconfiguration of electrical power networks, aircraft sequencing at an airport, timetabling, and scheduling (Van Roy 2005).

An example of using Mozart/Oz to seek admissible solutions is illustrated in Fig. 3.13. The search tree illustrates choice nodes as purple circles and solution nodes as green diamonds. The fully explored subtrees, which do not contain solution nodes, are presented as a single red triangle. There are 10 admissible solutions for the above-defined variables, their domains, and constraints.

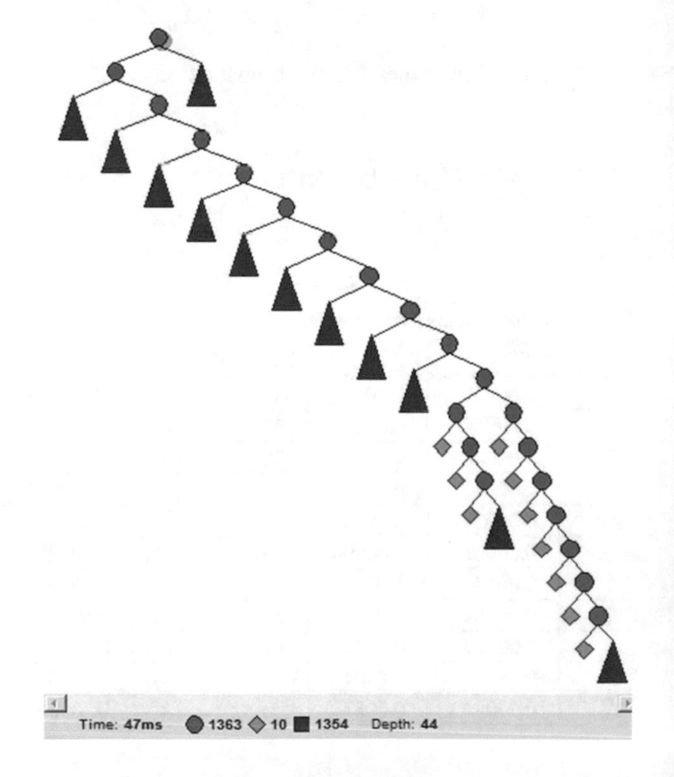

Fig. 3.13 The explored search tree for the basic variant

Table 3.5 A set of possible solutions

Case	Values of decision variables	Np
1	$X_9 = 78\%, X_{11} = 3, X_{12} = 4, X_{15} = 8, X_{17} = 3, X_{18} = 3$	440.2
2	$X_9 = 79\%, X_{11} = 3, X_{12} = 4, X_{15} = 8, X_{17} = 3, X_{18} = 3$	440.1
3	$X_9 = 80\%, X_{11} = 3, X_{12} = 4, X_{15} = 8, X_{17} = 3, X_{18} = 3$	440.0
4	$X_9 = 74\%, X_{11} = 3, X_{12} = 4, X_{15} = 8, X_{17} = 2, X_{18} = 3$	440.3
5	$X_9 = 75\%, X_{11} = 3, X_{12} = 4, X_{15} = 8, X_{17} = 2, X_{18} = 3$	440.2
6	$X_9 = 76\%, X_{11} = 3, X_{12} = 4, X_{15} = 8, X_{17} = 2, X_{18} = 3$	440.2
7	$X_9 = 77\%, X_{11} = 3, X_{12} = 4, X_{15} = 8, X_{17} = 2, X_{18} = 3$	440.1
8	$X_9 = 78\%, X_{11} = 3, X_{12} = 4, X_{15} = 8, X_{17} = 2, X_{18} = 3$	440.1
9	$X_9 = 79\%, X_{11} = 3, X_{12} = 4, X_{15} = 8, X_{17} = 2, X_{18} = 3$	440.0
10	$X_9 = 80\%, X_{11} = 3, X_{12} = 4, X_{15} = 8, X_{17} = 2, X_{18} = 3$	440.0

Table 3.5 presents 10 possible solutions of the above-described problem in terms of potential changes in decision variables and their impact on the desirable profit from a new product (Np).

Let us assume that the decision-maker's expectations for the profit increase from 440 to 450 thousand €. There is no solution satisfying a new value of the desirable profit, which triggers the next step in the procedure of solving the problem, i.e. verification of the possibility to extend domains related to the selected decision variables (see Fig. 3.12).

An Extension of Domains for the Selected Decision Variables

This step of the procedure of solving the problem stated in the inverse form requires selecting decision variables, for which an extension of domains is possible. As a result, the above-described problem is solved again for new domains assigned to some decision variables, and for other conditions unchanged. For example, domains for the following decision variables are extended: the number of employees involved in product design (X_{11} from 3#5 to 3#6), the duration of elaborating product design (X_{12} from 4#6 to 4#7), the number of employees involved in prototype tests (X_{17} from 2#3 to 2#4), and the duration of prototype tests (X_{18} from 3#5 to 3#6). Figure 3.14 illustrates the explored search tree generated in Mozart/Oz for the desirable profit at 450 thousand € and changed domains. Other domains of variables and constraints are the same as in the basic variant.

An extension of domains for four variables causes the increase of choice nodes. Nevertheless, the fully explored subtrees do not contain any solution nodes. If domains of decision variables cannot be extended or there is no solution (as in the presented example), then the possibility to change the set of decision variables is verified.

A Modification in the Set of Decision Variables

In the next step of the procedure of solving the problem stated in the inverse form, the decision-maker selects another set of decision variables towards reaching the desirable profit. The decision-maker can be supported in this task through the comparison of increments in the cost and sales volume (see Table 3.4). For example, to

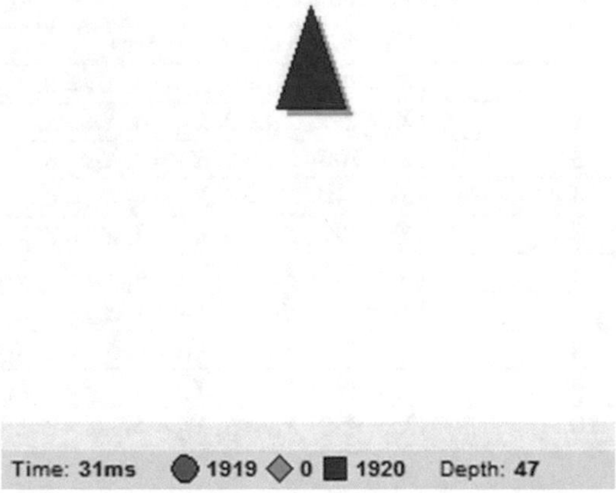

Fig. 3.14 The explored search tree for the extended domains

the present set of decision variables is added one variable more, namely the amount of materials needed to produce a unit of product (X_{21}). The planned materials consumption of producing a new product equals 6 units. The decision-maker is interested in obtaining information about changes in the total costs, sales, and profit, for materials consumption of producing a new product between 5 and 7 units. Domains for the decision variables are as follows: X_9 = 70%#80%, X_{11} = 3#6, X_{12} = 4#7, X_{15} = 8#10, X_{17} = 2#4, X_{18} = 3#6, and X_{21} = 5#7. The desirable profit should be greater than or equal to 450 thousand €. Other domains of variables and constraints are the same as in the basic variant. Figure 3.15 illustrates the explored search tree obtained in Mozart/Oz for the above-presented conditions.

There are four admissible solutions for materials consumption of producing a new product equalling 5 units. If the set of decision variables cannot be changed or there is no solution, then the possibility to change constraints is verified.

A Modification in the Set of Constraints

In the last step of the procedure of solving the problem stated in the inverse form, the decision-maker can change one or more constraints towards reaching the desirable profit. For example, the decision-maker wants to check the possibility of reaching the desirable profit through extending the duration of product design and prototype tests (in months) from 7 to 8 ($X_{12} + X_{18} \leq 8$). Other constraints, domains of variables, and the profit (450 thousand €) are the same as in the basic variant. Figure 3.16 illustrates the explored search tree generated in Mozart/Oz for these conditions.

Finally, if any constraints cannot be changed, then the last stop condition leads to the empty solution set.

Table 3.6 presents the results of seeking admissible solution for three cases and different strategies of variable distribution. First case refers to the basic variant, the second case to an extension of domains for the selected decision variables, and the third case to a modification in the set of decision variables. Different strategies of variable distribution in constraint programming are compared to exhaustive search (ES) in the context of the number of nodes checked, depth, and time needed to find

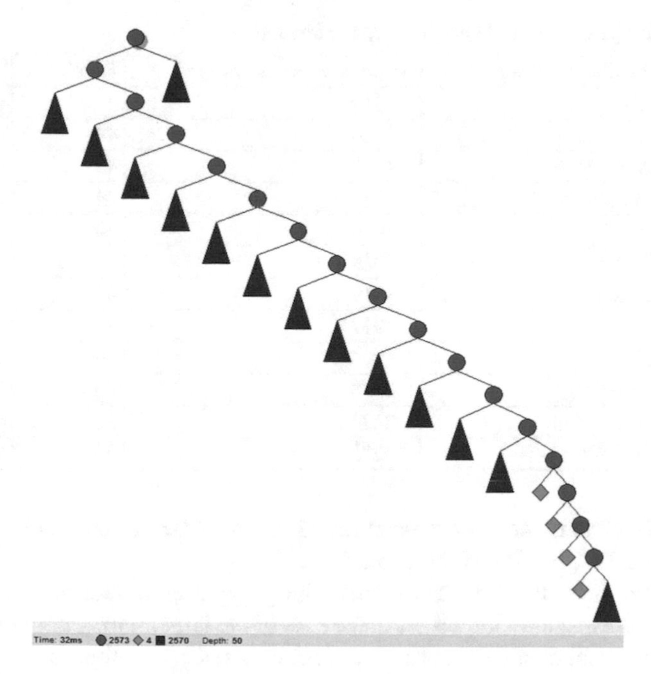

Fig. 3.15 The explored search tree for a new set of decision variables

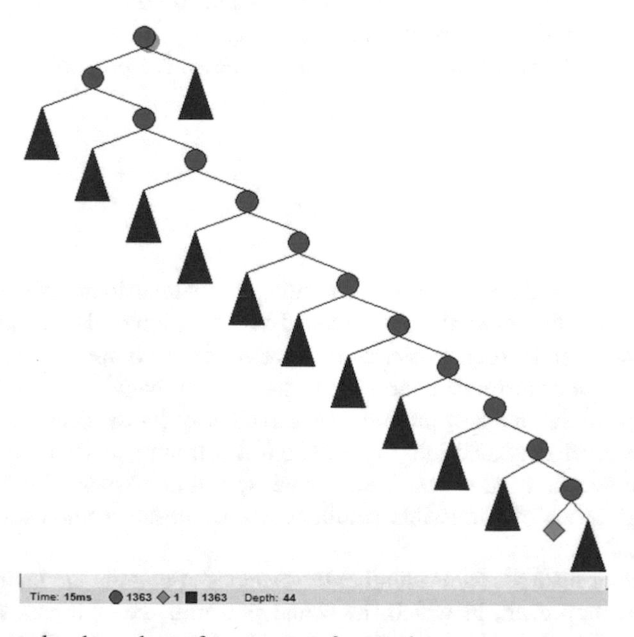

Fig. 3.16 The explored search tree for a new set of constraints

Table 3.6 Comparison of strategies for variable distribution

Case	Distribution strategy	Number of nodes checked	Depth	Time (s)
1	ES	3512	47	0.79
	CP naïve	1363	44	0.47
	CP first-fail	1363	44	0.31
	CP split	1363	44	0.22
2	ES	4096	54	1.14
	CP naïve	1919	47	0.31
	CP first-fail	1919	47	0.26
	CP Split	1919	47	0.21
3	ES	13,824	60	2.15
	CP naïve	2273	50	0.32
	CP first-fail	2273	50	0.22
	CP Split	2273	50	0.14

solutions. The calculations have been tested on an AMD Turion(tm) II Ultra Dual-Core M600 2.40 GHz, RAM 2 GB platform.

The results show that the use of constraint programming reduces computational time, which is especially important in the case of the larger number of possible solutions. The user can obtain the entire set of solutions or one optimal solution (e.g. for the maximal profit from a new product). For example in the first case, the user can obtain all admissible solutions (10 instances) or one optimal solution. Constraint programming enables the use of strategies related to constraint propagation and variable distribution, significantly reducing a set of admissible solutions and the average computational time, which improves interactive properties of a decision support system.

Summary

The proposed method aims to solve NPD-related problems in an effective way. The problem stated in the forward form is related to evaluation of the potential of a new product. This potential (e.g. measured by the cost, sales volume, or profit) is evaluated with the use of artificial neural networks and compared with multiple regression. If the potential of a new product is unsatisfactory for the decision-maker, then the problem is reformulated into the inverse form. It aims to search possible solutions within specified variables, their domains, and constraints. To search effectively a large space of admissible solutions, the proposed method uses constraint programming.

The existing methods for evaluating the product's potential are mainly based on an analogical approach, in which the actual performance of a past similar NPD project is used to estimate the performance (duration, cost, profit, etc.) of a new product. The proposed method develops a parametric approach that uses input and output variables to identify cause-and-effect relationships between these variables based on data from previous similar projects. The aim of using parametric estimation is twofold: the evaluation of the product's potential more precisely and the use of identified relationships to seek possible variants of the NPD project performance.

Enterprise databases contain numerous datasets, including the process of product development. This process involves many factors that can be interrelated and directly or indirectly affect the success of a new product. As relationships between the mentioned factors are often non-linear and incoherent, artificial neural networks are proposed to identify these relationships. The performed experiments show that artificial neural networks are able to estimate non-linear functions more precisely than multiple regression.

The problem stated in the inverse form is specified in terms of a CSP, including a set of variables, their domains, and constraints. A proposed framework for solving the problem stated in the inverse form consists of four stages of seeking possible solutions. These stages refer to verification of existing solutions for: (1) a set of decision variables within the specified input and output variables, their domains, and constraints; (2) extended domains related to the selected decision variables; (3) the modified set of decision variables; and (4) the modified set of constraints. The proposed approach supports the decision-maker in obtaining information of potentially more beneficial variants of product development. The performed experiments show that the use of constraint programming significantly reduces the search space and time needed to obtain results in comparison with the entire search space. This reduction is particularly important for problems specified in an inverse form, in which the increase in the number of decision variables and the extension of their domains usually result in the enormous number of potential solutions.

The potential of a new product is a crucial issue for selecting an optimal portfolio of NPD projects, and reallocating resources after incorporating a new project into the portfolio. It is proved that artificial neural networks can be successfully applied to the prediction. Consequently, their use to produce more precise estimates of the product's potential (compared to multiple regression) tends to improve decisions within the selection of an NPD project portfolio with the greatest potential and scheduling NPD projects. The proposed method also enables consideration of imprecise evaluations specified by experts through using fuzzy logic. Moreover, the use of neuro-fuzzy networks provides *if-then* rules based on imprecise values of factors affecting product development. As a result, the proposed method supports the decision-maker in obtaining reliable information about the product's potential, using enterprise databases and experts' opinions.

The proposed approach can be seen as methodology for continuous improvement of the NPD process towards identifying prerequisites to reach a desirable outcome of an NPD project. Thus, the decision-maker obtains information whether additional resources are able to switch the project towards desirable performance.

References

Apt, K. (2003). *Principles of constraint programming*. Cambridge: Cambridge University Press.

Awasthi, A., Grzybowska, K., Hussain, M., Chauhan, S. S., & Goyal, S. K. (2014). Investigating organizational characteristics for sustainable supply chain planning under fuzziness. In *Supply chain management under fuzziness* (pp. 81–100). Berlin: Springer.

Banaszak, Z., Zaremba, M., & Muszyński, W. (2009). Constraint programming for project-driven manufacturing. *International Journal of Production Economics, 120*, 463–475.

Bode, J., Ren, S., Luo, S., Shi, Z., Zhou, Z., Hu, H., Jiang, T., & Liu, B. (1995). Neural networks in new product development. In *Computer applications in production engineering* (pp. 659–666). Boston: Springer.

Bojadziev, G., & Bojadziev, M. (2007). Fuzzy logic for business, finance, and management. In *Advances in fuzzy systems: Applications and theory* (Vol. 23). Singapore: World Scientific.

Carlsson, C., Fedrizzi, M., & Fullér, R. (2004). *Fuzzy logic in management*. New York: Springer.

Castellano, G., Castiello, C., Fanelli, A. M., & Jain, L. (2007). Evolutionary neuro-fuzzy systems and applications. In L. Jain et al. (Eds.), *Advances in evolutionary computing for system design* (pp. 11–45). Berlin: Springer.

Cavalieri, S., Maccarrone, P., & Pinto, R. (2004). Parametric vs. neural network models for the estimation of production costs: A case study in the automotive industry. *International Journal of Production Economics, 91*(2), 165–177.

Czogala, E., & Leski, J. (2000). *Fuzzy and neuro-fuzzy intelligent systems*. Berlin: Springer.

Duran, O., Maciel, J., & Rodriguez, N. (2012). Comparisons between two types of neural networks for manufacturing cost estimation of piping elements. *Expert Systems with Applications, 39*(9), 7788–7795.

Eberhart, R. C., & Shi, Y. (2007). *Computational intelligence*. Burlington: Morgan Kaufmann.

Efendigil, T., Önüt, S., & Kahraman, C. (2009). A decision support system for demand forecasting with artificial neural networks and neuro-fuzzy models: A comparative analysis. *Expert Systems with Applications, 36*(3), 6697–6707.

Engelbrecht, A. P. (2007). *Computational intelligence: An introduction*. Chichester: Wiley.

Fazlollahtabar, H., & Mahdavi-Amiri, N. (2013). Design of a neuro-fuzzy–regression expert system to estimate cost in a flexible jobshop automated manufacturing system. *The International Journal of Advanced Manufacturing Technology, 67*(5-8), 1809–1823.

Finnie, G. R., Wittig, G. E., & Desharnais, J. M. (1997). A comparison of software effort estimation techniques: Using function points with neural networks, case-based reasoning and regression models. *Journal of Systems and Software, 39*(3), 281–289.

Frühwirth, T., & Abdennadher, S. (2003). *Essentials of constraint programming*. Berlin: Springer.

Gil-Lafuente, A. M. (2005). *Fuzzy logic in financial analysis*. Berlin: Springer.

Gumus, A. T., Guneri, A. F., & Keles, S. (2009). Supply chain network design using an integrated neuro-fuzzy and MILP approach: A comparative design study. *Expert Systems with Applications, 36*(10), 12570–12577.

Huang, H. Z., Bo, R., & Chen, W. (2006). An integrated computational intelligence approach to product concept generation and evaluation. *Mechanism and Machine Theory, 41*(5), 567–583.

Hudec, M. (2016). *Fuzziness in information systems: How to deal with crisp and fuzzy data in selection, classification, and summarization*. Berlin: Springer.

Kar, S., Das, S., & Ghosh, P. K. (2014). Applications of neuro fuzzy systems: A brief review and future outline. *Applied Soft Computing, 15*, 243–259.

Kim, G. H., An, S. H., & Kang, K. I. (2004). Comparison of construction cost estimating models based on regression analysis, neural networks, and case-based reasoning. *Building and Environment, 39*(10), 1235–1242.

Konar, A. (2006). *Computational intelligence: Principles, techniques and applications*. Berlin: Springer Science & Business Media.

Kwong, C. K., Wong, T. C., & Chan, K. Y. (2009). A methodology of generating customer satisfaction models for new product development using a neuro-fuzzy approach. *Expert Systems with Applications, 36*(8), 11262–11270.

Latif, H. H., Paul, S. K., & Azeem, A. (2014). Ordering policy in a supply chain with adaptive neuro-fuzzy inference system demand forecasting. *International Journal of Management Science and Engineering Management, 9*(2), 114–124.

Lee, H., Kim, S. G., Park, H. W., & Kang, P. (2014). Pre-launch new product demand forecasting using the Bass model: A statistical and machine learning-based approach. *Technological Forecasting and Social Change, 86*, 49–64.

Li, S. (2000). The development of a hybrid intelligent system for developing marketing strategy. *Decision Support Systems, 27*(4), 395–409.

Medsker, L. R. (2012). *Hybrid intelligent systems*. New York: Springer Science & Business Media.

Mitra, S., & Hayashi, Y. (2000). Neuro-fuzzy rule generation: survey in soft computing framework. *IEEE Transactions on Neural Networks, 11*(3), 748–768.

Narver, J. C., Slater, S. F., & MacLachlan, D. L. (2004). Responsive and proactive market orientation and new-product success. *Journal of Product Innovation Management, 21*, 334–347.

Nauck, D. D., & Nürnberger, A. (2013). Neuro-fuzzy systems: A short historical review. In C. Moewes et al. (Eds.), *Computational intelligence in intelligent data analysis* (pp. 91–109). Berlin: Springer.

Paliwal, M., & Kumar, U. A. (2009). Neural networks and statistical techniques: A review of applications. *Expert Systems with Applications, 36*, 2–17.

Pedrycz, W. (2006). A quest for granular computing and logic processing. In *Advances in computational intelligence: theory & applications*. World Scientific Publishing.

Prieto, A., Prieto, B., Ortigosa, E. M., Ros, E., Pelayo, F., Ortega, J., & Rojas, I. (2016). Neural networks: An overview of early research, current frameworks and new challenges. *Neurocomputing, 214*, 242–268.

Rajab, S., & Sharma, V. (2018). A review on the applications of neuro-fuzzy systems in business. *Artificial Intelligence Review, 49*(4), 481–510.

Relich, M. (2008). The using of fuzzy-neural system to monitoring and control of liquidity in a small business. *Management, 12*(1), 295–305.

Relich, M. (2010). A decision support system for alternative project choice based on fuzzy neural networks. *Management and Production Engineering Review, 1*(4), 46–54.

Relich, M. (2012). An evaluation of project completion with application of fuzzy set theory. *Management, 16*(1), 216–229.

Relich, M. (2016). Computational intelligence for estimating cost of new product development. *Foundations of Management, 8*(1), 21–34.

Relich, M., & Bzdyra, K. (2015). Knowledge discovery in enterprise databases for forecasting new product success. In *International Conference on Intelligent Data Engineering and Automated Learning* (pp. 121–129). Cham: Springer.

Relich, M., & Muszyński, W. (2014). The use of intelligent systems for planning and scheduling of product development projects. *Procedia Computer Science, 35*, 1586–1595.

Relich, M., & Pawlewski, P. (2017). A fuzzy weighted average approach for selecting portfolio of new product development projects. *Neurocomputing, 231*, 19–27.

Relich, M., & Pawlewski, P. (2018). A case-based reasoning approach to cost estimation of new product development. *Neurocomputing, 272*, 40–45.

Rossi, F., Van Beek, P., & Walsh, T. (2006). *Handbook of constraint programming*. New York: Elsevier Science.

Rutkowska, D. (2002). *Neuro-fuzzy architectures and hybrid learning*. Berlin: Springer.

Seo, K. K., Park, J. H., Jang, D. S., & Wallace, D. (2002). Approximate estimation of the product life cycle cost using artificial neural networks in conceptual design. *The International Journal of Advanced Manufacturing Technology, 19*, 461–471.

Seyedhoseini, S. M., Jassbi, J., & Pilevari, N. (2010). Application of adaptive neuro fuzzy inference system in measurement of supply chain agility: Real case study of a manufacturing company. *African Journal of Business Management, 4*(1), 83–96.

Siddique, N., & Adeli, H. (2013). *Computational intelligence: Synergies of fuzzy logic, neural networks and evolutionary computing*. New York: Wiley.

Smith, A. E., & Mason, A. K. (1997). Cost estimation predictive modeling: Regression versus neural network. *The Engineering Economist, 42*(2), 137–161.

Song, X. M., & Parry, M. E. (1997). A cross-national comparative study of new product development processes: Japan and the United States. *The Journal of Marketing, 61*, 1–18.

Suthaharan, S. (2016). *Machine learning models and algorithms for big data classification*. Boston: Springer.

Tkac, M., & Verner, R. (2016). Artificial neural networks in business: Two decades of research. *Applied Soft Computing, 38*, 788–804.

Van Roy, P. (2005). Multiparadigm programming in Mozart/Oz. In *Lecture notes in computer science* (Vol. 3389). Berlin: Springer.

Van Roy, P., & Haridi, S. (2004). *Concepts, techniques, and models of computer programming*. Cambridge: Massachusetts Institute of Technology.

Wang, L., & Fu, X. (2006). *Data mining with computational intelligence*. Berlin: Springer.

Wang, F. K., Chang, K. K., & Tzeng, C. W. (2011). Using adaptive network-based fuzzy inference system to forecast automobile sales. *Expert Systems with Applications, 38*(8), 10587–10593.

Zadeh, L. A. (1965). Information and control. *Fuzzy Sets, 8*(3), 338–353.

Zemouri, R., Gouriveau, R., & Zerhouni, N. (2010). Defining and applying prediction performance metrics on a recurrent NARX time series model. *Neurocomputing, 73*, 2506–2521.

Zhang, G. P. (2010). Neural networks for data mining. In O. Maimon & L. Rokach (Eds.), *Data mining and knowledge discovery handbook* (pp. 419–444). New York: Springer.

Chapter 4
A Decision Support System for Portfolio Management of NPD Projects

Decision Support Systems for Product Development

Decisions Involved in Product Development

Product development begins from market analysis and defining a new product according to a company's strategy. In this phase, decisions refer to product's characteristics (in the context of unique features of the product, product design and engineering, innovative technologies, production processes, etc.) and product positioning. These strategic decisions release further decisions in successive NPD phases, i.e. idea generation, concept selection, design, and tests of the most promising concepts, taking into account the optimal cost and time of an NPD project and new product quality. Uncertainties in this NPD phase are related to competitors' choices, customers' preferences, and the company's access to required resources to achieve a desirable product.

The idea generation phase and concept evaluation phase are closely related. In the idea generation phase, decisions refer to the selection of team members and techniques for generating new product ideas and a procedure for detecting potentially valuable ideas. The wrong selection of ideas can result in developing unsuccessful products or rejecting ideas that could be successfully developed. In the concept evaluation phase, selected ideas are further specified into concepts that can be considered in the context of required materials, labour, technology, and related costs. These factors are compared with expected profits in order to identify the most promising concepts for engineering design.

In the product design phase, the R&D team makes decisions related to engineering, prototype tests, and relevant changes in the production process. Moreover, decisions in this phase are closely related to project management processes in the field of planning, monitoring, and controlling project scope, schedule, cost, quality, resources, communication, risk, and procurement. Design decisions involve a

M. Relich, *Decision Support for Product Development*, Computational Intelligence Methods and Applications,
https://doi.org/10.1007/978-3-030-43897-5_4

detailed product definition, including its shape, sizes, components, materials, reliability, etc. New technologies or materials used in the product can impose adjustment of the production process and logistic system. The product design phase contains uncertainties regarding acquisition of required technologies and materials, development of a new product according to customers' needs, and homologation of the product. Moreover, uncertainties are related to completion of an NPD project within the acceptable time (before competitors), cost, and quality.

Decisions in the commercialisation phase refer to the time and budget of advertising campaign, service associated with the product, and the production planning. Uncertainties in this phase are related to customers' acceptance, competitors' responses, and partners' decisions (e.g. within a service system supporting the product and/or customer). The level of uncertainties can be reduced with the use of structured actions towards understanding market responses to a new product and benchmarking (Montagna 2011).

A Framework for Decision Support Systems

A highly competitive environment forces companies to improve the effectiveness of the NPD process. As new products should be closer to customers' needs, and have quicker time to market and higher reliability, companies are looking for new forms of cooperation and coordination among their employees and customers. Consequently, the NPD process requires dedicated tools and methods for acquiring, formalising, and using essential information for pattern recognition, and supporting users in decision-making. The use of decision support systems provides an opportunity for managers and organisations to reduce their efforts within more effective and agile decisional processes.

Decision support systems (DSS) are interactive computer-based systems that help decision-makers utilise data and models to solve unstructured problems (Sprague and Carlson 1982). The decision-making process in terms of DSS design includes problem recognition and problem definition to facilitate the creation of models, alternative generation and model development to analyse alternatives, and finally, selection and presentation of the best solution to users. Typical DSSs consist of three components dedicated to (Shim et al. 2002):

- Sophisticated database management capabilities with access to internal and external data, information, and knowledge
- Modelling functions provided by a model management subsystem
- User interface that enables interactive queries and reporting

Figure 4.1 illustrates main components included in a decision support system.

Internal data refer to past NPD projects, production capability, resource availability, etc., and it can be retrieved from an enterprise system. In turn, external data can contain market trends, including customers' expectations for new products, and it is acquired from clients, suppliers, or competitors. A database management

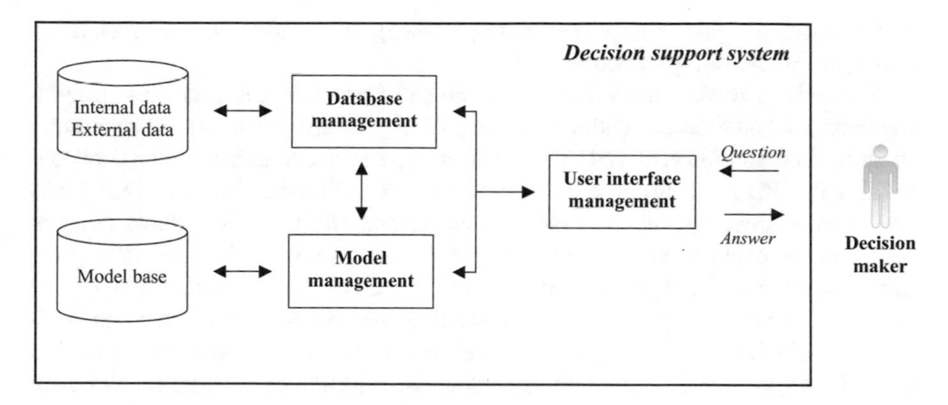

Fig. 4.1 Components of a decision support system

subsystem retrieves data that can be used in problem-solving, and supplies it to a model management subsystem that enables pattern identification among data, as well as generation and evaluation of alternatives using models specified in a model base. These models are related to statistical, optimisation, what-if methods, etc., which are specified according to a decision problem (e.g. for project portfolio selection). A user interface management should allow the decision-maker to identify the best alternative(s) for problem solution and check results for changed parameters.

The Use of DSS in Different Product Development Phases

Product development involves strategic decisions referring to mission approval (including the selection of product lines), the selection of the most promising concepts for an NPD project portfolio, the desirable quality of new products, production approval, and finally the time of launch (see Fig. 1.6). In product development, DSSs are also used in operational decisions, such as the selection of optimal configuration design, optimal production schedules, and optimal advertising expenses.

DSSs are rarely used in the first phase of the NPD process, namely in market analysis. Matsatsinis and Siskos (2003) present DSSs for predicting customer purchasing behaviour, market segmentation, and selection of penetration strategy of the product, using product, business, and market-related data (e.g. price of the product, sales volume, customer's decision-making pattern). Hallstedt et al. (2010) consider decision systems for strategic capability assessment from a sustainability perspective, using the following data: life cycle assessment of product projects, customers' requirements, and prioritisations from the senior management. Chan and Ip (2011) propose a dynamic decision support system for predicting the value of customer and selecting the most competitive products to launch, using data related to products (performance, quality, design, packaging, competitiveness), customers (satisfaction, weights of requirement importance), and marketing strategies (marketing and remarketing effectiveness).

Similarly to market analysis, the next phase of product development—generating concepts and their evaluation towards project portfolio selection—is also rarely specified in DSSs. Oh et al. (2012) propose an expert system for improving decision-making in NPD portfolio management on the basis of financial indices (NPV, revenue, sales, cost, quantity), strategic importance (fit with key initiatives and priorities, innovation relative to market), commercial potential (base NPV, gross profit margin, user base growth), technical risk, and commercial risk. Relich (2013) proposes decision support tools for evaluating the cost and duration of an NPD project. Achiche et al. (2013) present decision support tools in the core front-end activities of product development, including the selection of product development team and estimation of investment cost, and using person-related factors (working hours, training, professional background) and costs (material, staff, and time related to an NPD project). Relich (2015) proposes a knowledge-based system for predicting new product success that uses selected computational intelligence techniques to identify relationships between parameters related to a new product and product success. A knowledge-based system for new product portfolio selection in the context of using enterprise databases is presented by Relich (2016). In turn, Mirtalaie et al. (2017) propose a decision support framework for identifying novel ideas in new product development, using features and/or attributes of products, and customers' opinions about proposed features.

Compared to market analysis and concept generation and evaluation, the use of DSSs in product design is more frequent among the researchers. Besharati et al. (2006) propose a decision support system for product design, selecting a design alternative according to the designer's utility on the basis of market demand, market segments, product attributes (performance and market related), and manufacturing cost. Gandy et al. (2007) present decision support for evaluating product concepts based on design requirements (dimensions, material) and lifetime of a product. Abramovici and Lindner (2013) propose knowledge-based decision support for improving standard products through product design change, product improvement prototyping, and improved product operation, monitoring, and evaluation. They use data related to customers, products (technical data, service data, product use data), and external improvement drivers (data about similar competitive products, new relevant technologies, laws and regulations). Lei and Moon (2015) present a decision support system for market-driven product positioning and design. Their DSS evaluates market segments for different property combinations of a new product and selects design parameters. In turn, Relich and Pawlewski (2016) propose a knowledge-based system for estimating cost of product design.

The use of a DSS for the commercialisation and market analysis phase is presented by Ching-Chin et al. (2010). Their DSS selects the best-suited forecasting model for data and forecasts new product sales on the basis of actual data points, product lifecycle length, and demand pattern. Moreover, Yang et al. (2016) propose a DSS for predicting consumer preference in NPD.

The Proposed Decision Support System

The proposed decision support system aims to support decision-makers in three NPD-related problems: evaluating the product's potential, project portfolio selection, and resource allocation. Figure 4.2 presents the proposed DSS in the context of questions and answers obtained.

The proposed decision support system for solving NPD-related problems (further called DSS4NPD) consists of the following parts:

- Specification of an NPD model (Fig. 4.3)
- Specification of values for input variables (Fig. 4.4)
- Evaluation of the product's potential (Fig. 4.5)
- Evaluation of the product's potential—ANN (Fig. 4.6)
- Evaluation of the product's potential: an inverse approach (Fig. 4.7)
- Evaluation of the product's potential—ANFIS (Fig. 4.8)
- Project portfolio selection (Fig. 4.9)
- Project portfolio selection: an inverse approach (Fig. 4.10)
- Project scheduling (Fig. 4.11)
- Project scheduling: an inverse approach (Fig. 4.12)

Figure 4.3 illustrates specification of an NPD model in DSS4NPD, including import data (input and output variables related to past NPD projects) from enterprise databases, selection of input and output variables for analysis, and specification of constraints. The user can also save data for further analysis using

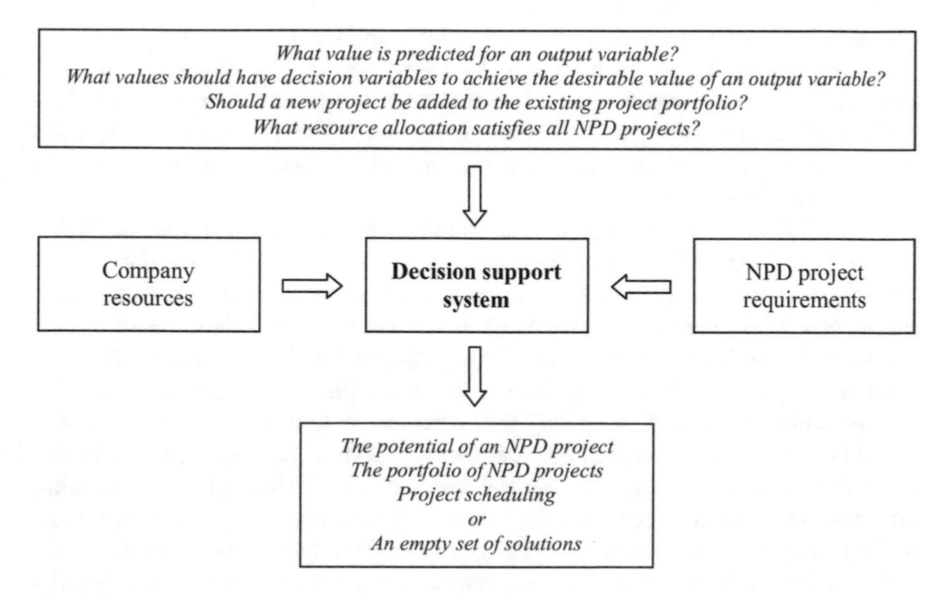

Fig. 4.2 The proposed decision support system

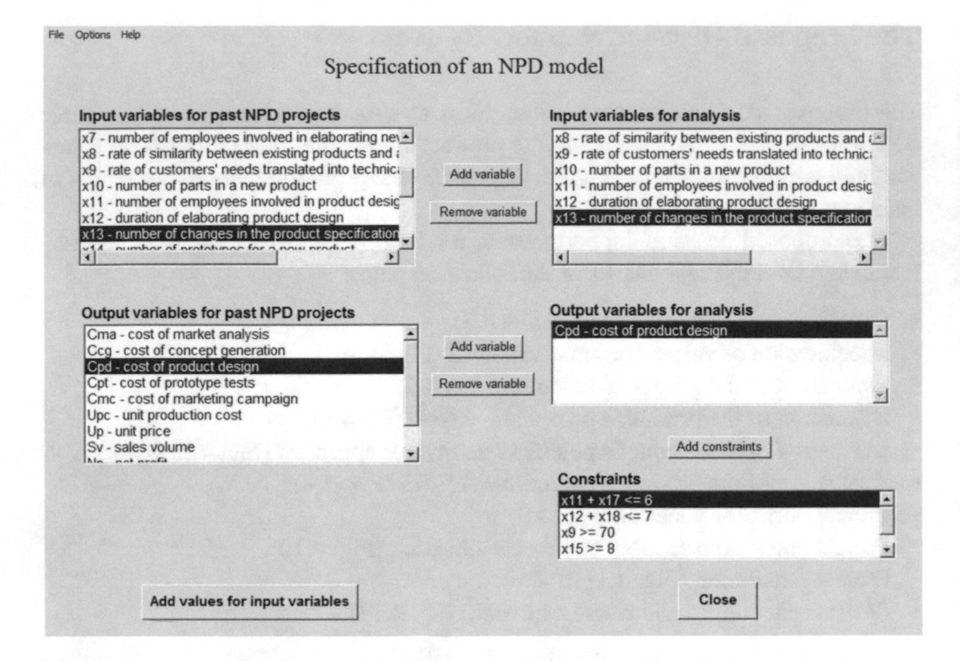

Fig. 4.3 DSS4NPD—Specification of an NPD model

DSS4NPD. After specifying variables and constraints, the user adds the expected values for input variables, including their domains.

Figure 4.4 illustrates specification of values for previously selected input variables. DSS4NPD allows the user to specify values for input variables and their domains related to a new project and select the NPD-related problem to further analysis.

After selecting the problem referred to as evaluation of the product's potential, DSS4NPD offers two options regarding the use of a database related to past NPD projects and the use of experts' opinions (see Fig. 4.5). The first option refers to the data specified in a precise form, whereas in the second option information can be input by the experts in an imprecise form.

Figure 4.6 illustrates evaluation of the product's potential for data specified in the precise form. DSS4NPD allows the user to obtain the predicted value of an output variable calculated with the use of an estimation method that generates the least mean absolute percentage error (MAPE). Errors are calculated for three estimation methods: parametric estimation using ANN, parametric estimation using MRA, and analogical approach. Moreover, there is the possibility of computing results for default values of an ANN or specify parameters to build and train ANN. The user can choose the type of a neural network (ANN GDM or ANN LM) and determine the optimal number of hidden layers and neurons in each hidden layer for training an ANN. The identified relationships can be saved to knowledge base for further analysis (e.g. project portfolio selection), or used for simulations within the problem stated in an inverse form (i.e. the search of admissible solutions, for which the desirable value of an output variable is achieved).

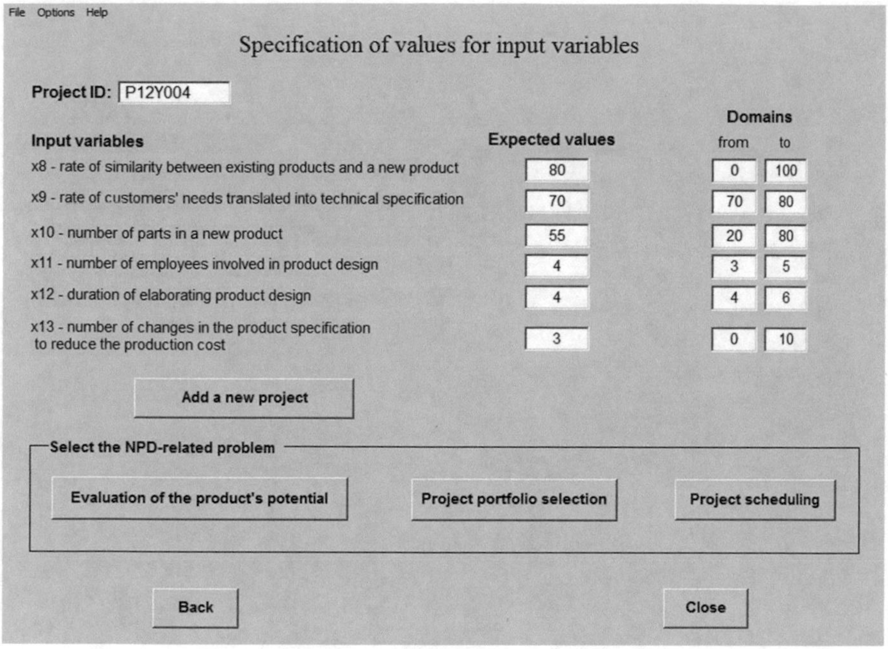

Fig. 4.4 DSS4NPD—Specification of values for input variables

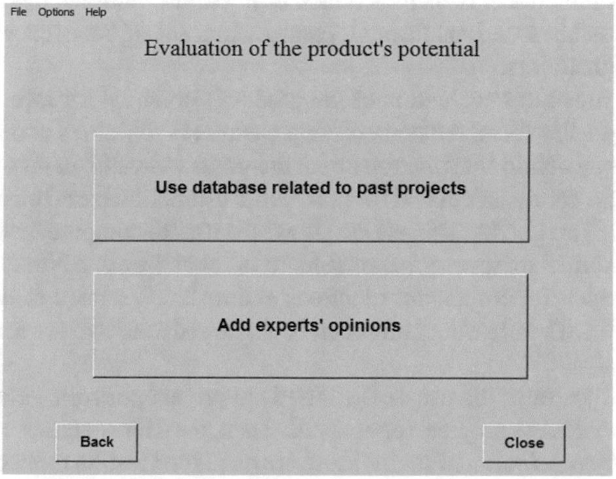

Fig. 4.5 DSS4NPD—Evaluation of the product's potential

Figure 4.7 illustrates DSS4NPD for the problem of evaluating the product's potential stated in an inverse form. The user specifies constraints related to the desirable value of an output variable (e.g. the net profit from an NPD project should not be less than 440 thousands € in the presented example) and selects decision

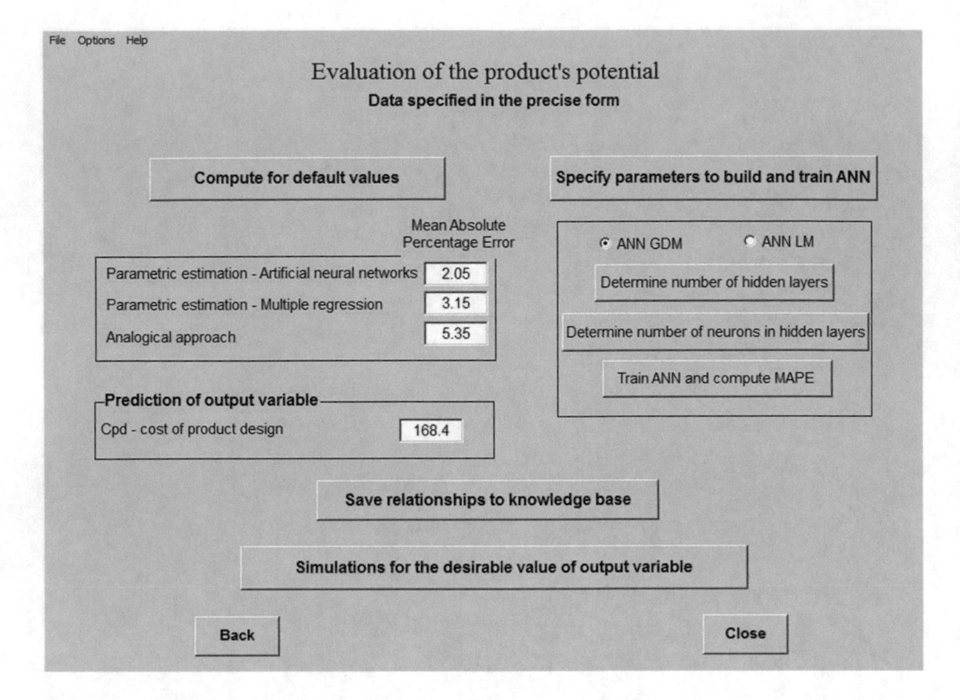

Fig. 4.6 DSS4NPD—Evaluation of the product's potential—ANN

variables for simulations. If there is a lack of possible solutions, then the user may try to seek possible solutions through changes in a set of decision variables, their domains, or constraints.

Figure 4.8 illustrates evaluation of the product's potential for information specified in the imprecise form. After specifying customers' opinions about a new product, the user may obtain information about the predicted value of an output variable and MAPEs. Errors are calculated for parametric estimation based on a fuzzy-neural system (ANFIS) and MRA. DSS4NPD allows the user to compute results for default values of an ANFIS or specify parameters to build and train ANFIS. The user can choose the shape of membership functions and methods related to defuzzification and train ANFIS. The identified rules can be displayed and saved to knowledge base for further analysis.

Figure 4.9 illustrates the use of DSS4NPD in project portfolio selection. Firstly, there are selected NPD projects for analysis. Then, the user specifies constraints and criteria for project selection. Finally, there are displayed results related to a set of an optimal project portfolio, total net profit, and NPD costs. The user may also search prerequisites towards reaching the desirable performance (total net profit) of the selected project portfolio.

Figure 4.10 illustrates DSS4NPD for the problem of project portfolio selection stated in an inverse form. The user specifies the desirable outcome (e.g. the total net profit from the NPD portfolio should not be less than 10,900 thousands €) and

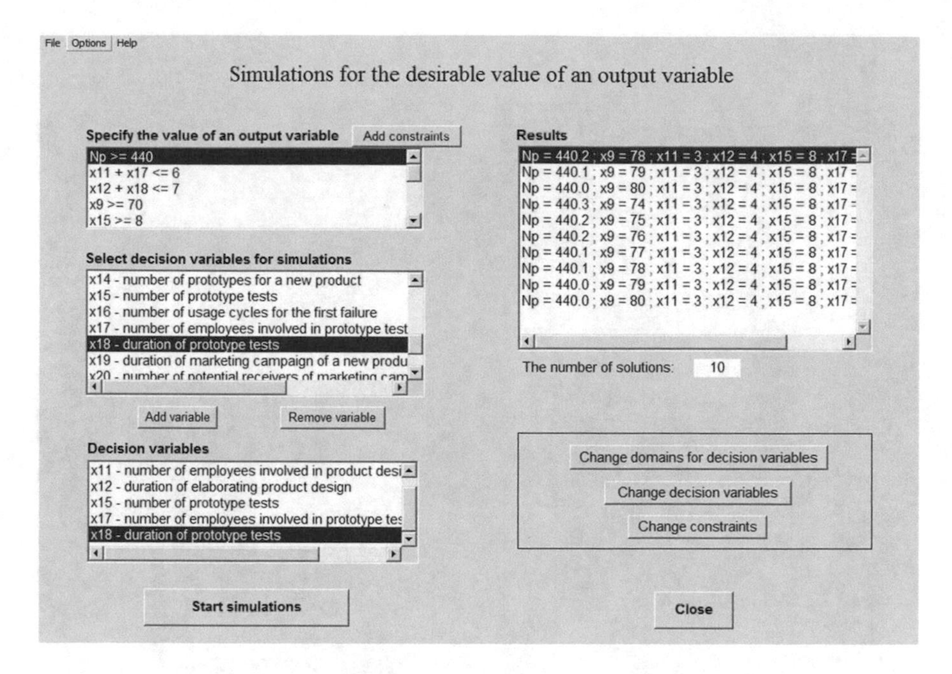

Fig. 4.7 DSS4NPD—Evaluation of the product's potential: an inverse approach

selects decision variables for simulations. If there is no possible solution, then the user may also check whether there are any solutions for changes in a set of decision variables, their domains, or constraints.

Figure 4.11 illustrates the use of DSS4NPD in project scheduling. The user selects NPD projects for scheduling and specifies activities for projects and relations between these activities in the form of precedence constraints. DSS4NPD allows the user to display schedule for selected projects and search resource allocation for the desirable completion time of a project portfolio.

Figure 4.12 illustrates DSS4NPD for the problem of project scheduling stated in an inverse form. The user specifies the desirable outcome (e.g. the maximal duration of project portfolio duration should not exceed 15 weeks) and selects decision variables for simulations. For example, DSS4NPD verifies the possibility of reducing project portfolio duration through increasing the number of employees involved in product design. If there is no possible solution, then the user may check whether there are any solutions changing a set of decision variables, their domains, or constraints. The changes in domains related to decision variables may increase the granularity of values, reducing the number of possible solutions (if needed).

The proposed DSS allows the decision-maker to specify an NPD model, identify relationships between input and output variables, estimate an output variable, and simulate effects of managerial decisions concerning project portfolio selection and resource allocation. Cause-and-effect relationships are identified on the basis of previous NPD projects and used for parametric estimation and simulation modelling.

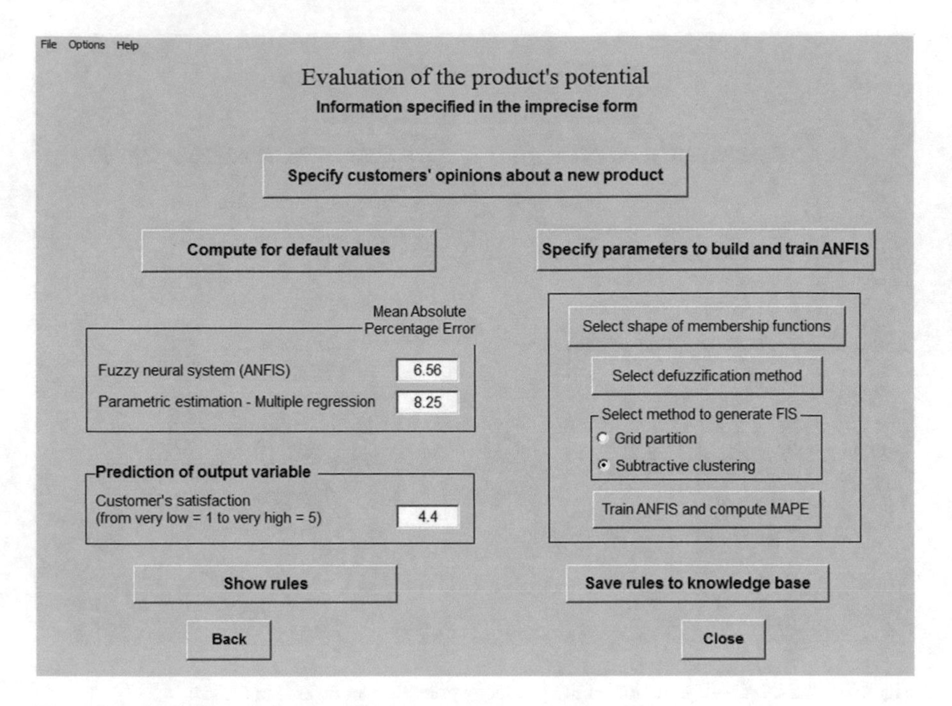

Fig. 4.8 DSS4NPD—Evaluation of the product's potential—ANFIS

The usefulness of the proposed approach has been verified in the group of Polish and Slovak enterprises that design components in the automotive industry. Verification results show that the use of simulation modelling in terms of declarative representation widens the scope of variants for alternative project executions compared to traditional scenario analysis that contains a basic, optimistic, and pessimistic scenario. Moreover, the use of parametric estimation based on some computational intelligence techniques increases the accuracy of cost estimation on average of 20% compared to multiple linear regression models. The incorporation of non-financial criteria into the evaluation of the product's potential (e.g. customers' opinions about their satisfaction with product design) improves the estimation accuracy on average of 25%. Moreover, the proposed DSS can be adapted to needs of a specific NPD project without changes in the structure of an enterprise information system, i.e. there are no requirements for changes in parameters within an enterprise system.

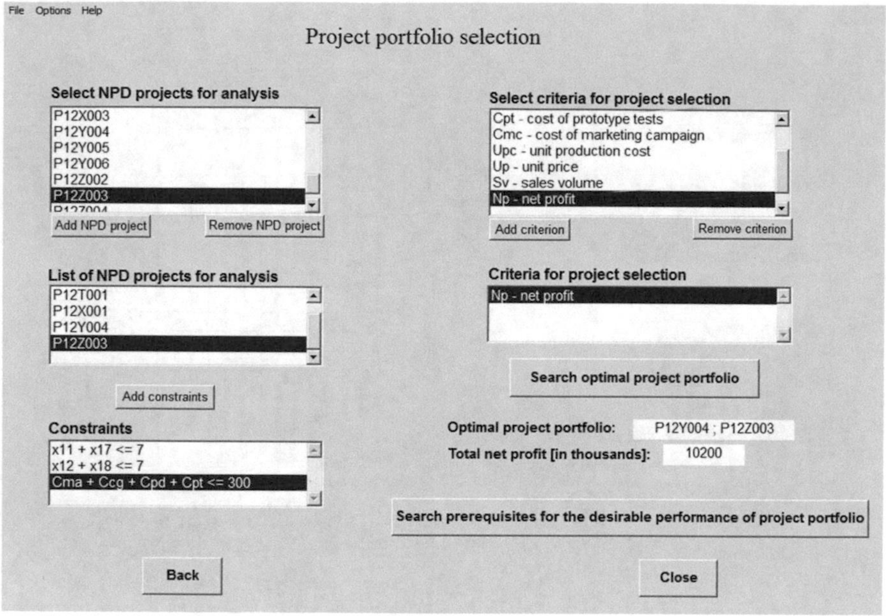

File Options Help
Project portfolio selection
Select NPD projects for analysis
P12X003
P12Y004
P12Y005
P12Y006
P12Z002
P12Z003
Add NPD project
Remove NPD project
Select criteria for project selection
Cpt - cost of prototype tests
Cmc - cost of marketing campaign
Upc - unit production cost
Up - unit price
Sv - sales volume
Np - net profit
Add criterion
Remove criterion
List of NPD projects for analysis
P12T001
P12X001
P12Y004
P12Z003
Criteria for project selection
Np - net profit
Search optimal project portfolio
Add constraints
Constraints
x11 + x17 <= 7
x12 + x18 <= 7
Cma + Ccg + Cpd + Cpt <= 300
Optimal project portfolio:
P12Y004 ; P12Z003
Total net profit [in thousands]:
10200
Search prerequisites for the desirable performance of project portfolio
Back
Close

Fig. 4.9 DSS4NPD—Project portfolio selection

File Options Help
Simulations for the desirable performance of project portfolio
Specify the desirable outcome
Add constraints
Np >= 10900
x11 + x17 <= 7
x12 + x18 <= 7
Cma + Ccg + Cpd + Cpt <= 300
x9 >= 70
Results
P12Y004 [x10 = 36 ; x15 = 12 ; x21 = 5] P12Z003 [x10 = 32 ; x1
P12Y004 [x10 = 37 ; x15 = 12 ; x21 = 5] P12Z003 [x10 = 32 ; x1
P12Y004 [x10 = 36 ; x15 = 12 ; x21 = 5] P12Z003 [x10 = 33 ; x1
P12Y004 [x10 = 37 ; x15 = 12 ; x21 = 5] P12Z003 [x10 = 33 ; x1
P12Y004 [x10 = 36 ; x15 = 12 ; x21 = 5] P12Z003 [x10 = 32 ; x1
Select decision variables for simulations
x16 - number of usage cycles for the first failure
x17 - number of employees involved in prototype test
x18 - duration of prototype tests
x19 - duration of marketing campaign of a new produ
x20 - number of potential receivers of marketing cam
x21 - amount of materials needed to produce a unit o
The number of solutions:
5
Add variable
Remove variable
Decision variables
x10 - number of parts in a new product
x15 - number of prototype tests
x21 - amount of materials needed to produce a unit o
Change domains for decision variables
Change decision variables
Change constraints
Start simulations
Close

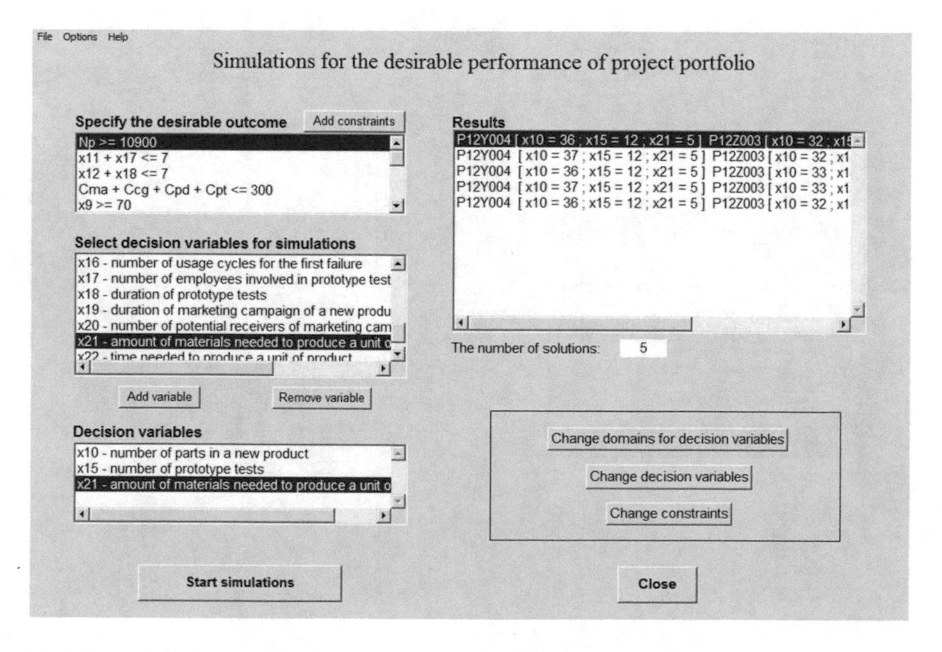

Fig. 4.10 DSS4NPD—Project portfolio selection: an inverse approach

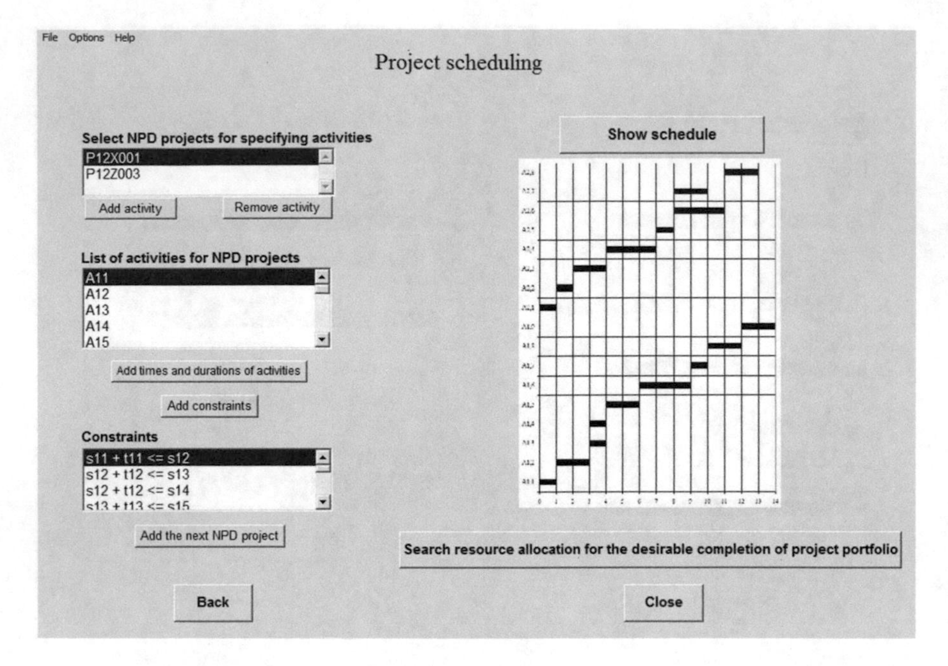

Fig. 4.11 DSS4NPD—Project scheduling

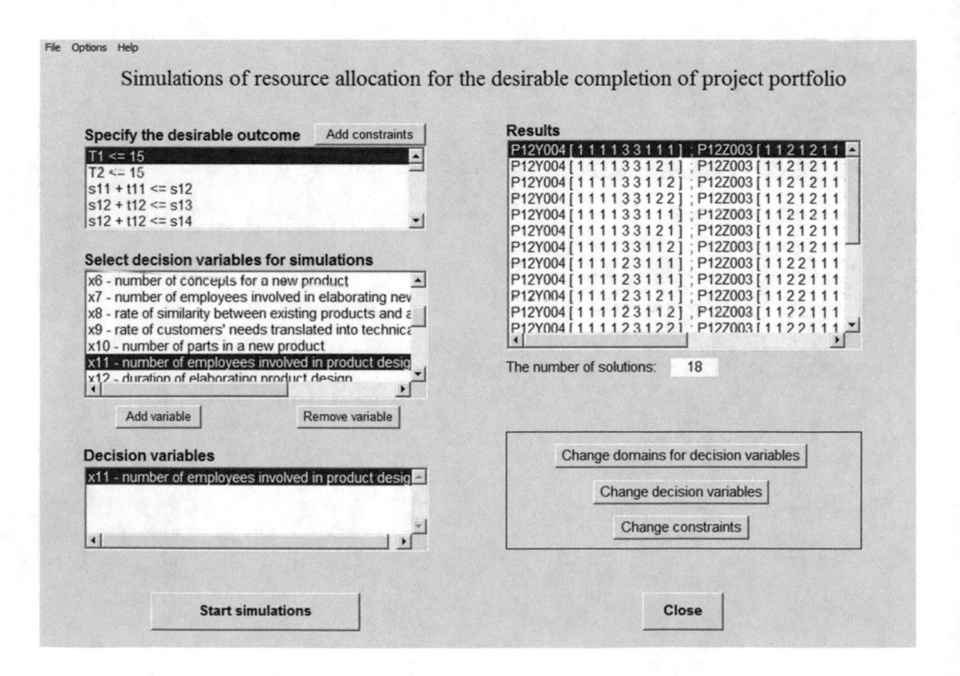

Fig. 4.12 DSS4NPD—Project scheduling: an inverse approach

Summary

Decision support systems aim to support managers in analytical phases of the decision-making process, i.e. in data acquisition, pattern recognition, and principally in creating decision variants. The ability of DSS to generate scenarios (decision variants along with prediction of their consequences) distinguishes DSS among other analytical information systems. Predictive abilities of DSS ensure their usefulness as a tool in decision analysis. Nevertheless, the usefulness of a DSS depends on a conceptual framework for the DSS design that includes goals and tasks of a DSS in an enterprise, structural and functional characteristic of a DSS, and principles for retrieving, processing, saving, and storing data.

This chapter presents decisions involved in product development, foundations for designing DSS, current applications of DSS in different phases of product development, and the proposed decision support system. DSS4NPD is dedicated to three NPD-related problems: evaluation of the new product's potential, project portfolio selection, and project scheduling. DSS4NPD has been developed according to the described methodology, including model specification in terms of CSP (variables, their domains, and constraints), and the application of computational intelligence techniques to parametric estimation of variables related to evaluation of the potential of a new product.

The use of CSP to model specification ensures adaptability of the NPD model to new conditions related to the changeable business environment (e.g. customers' opinions about the product's potential, accessible resources). Moreover, the specific problem can be stated in a forward and inverse form alternately. As a result, the decision-maker obtains information about the needed changes in an NPD project to reach the desirable value of an output variable (e.g. the NPD cost or net profit). The application of DSS4NPD has been verified in a real enterprise that fulfils requirements related to project management standards for collecting data and monitoring performance of NPD projects. The conducted simulations confirm the usefulness of constraint programming to reduce the search space of admissible solutions and the corresponding time, especially for solving the resource-constrained multi-project scheduling problem stated in an inverse form. The reduction of the search space is particularly significant not only from the perspective of computational performance but also from the perspective of managers' cognitive processes. A few simulations of economic effects of product development are easier to analyse than hundreds of them.

References

Abramovici, M., & Lindner, A. (2013). Knowledge-based decision support for the improvement of standard products. *CIRP Annals-Manufacturing Technology, 62*(1), 159–162.

Achiche, S., Appio, F. P., McAloone, T. C., & Di Minin, A. (2013). Fuzzy decision support for tools selection in the core front end activities of new product development. *Research in Engineering Design, 24*(1), 1–18.

Besharati, B., Azarm, S., & Kannan, P. K. (2006). A decision support system for product design selection: A generalized purchase modeling approach. *Decision Support Systems, 42*(1), 333–350.

Chan, S. L., & Ip, W. H. (2011). A dynamic decision support system to predict the value of customer for new product development. *Decision Support Systems, 52*(1), 178–188.

Ching-Chin, C., Ieng, A. I. K., Ling-Ling, W., & Ling-Chieh, K. (2010). Designing a decision-support system for new product sales forecasting. *Expert Systems with Applications, 37*(2), 1654–1665.

Gandy, A., Jäger, P., Bertsche, B., & Jensen, U. (2007). Decision support in early development phases – A case study from machine engineering. *Reliability Engineering & System Safety, 92*(7), 921–929.

Hallstedt, S., Ny, H., Robèrt, K. H., & Broman, G. (2010). An approach to assessing sustainability integration in strategic decision systems for product development. *Journal of Cleaner Production, 18*(8), 703–712.

Lei, N., & Moon, S. K. (2015). A decision support system for market-driven product positioning and design. *Decision Support Systems, 69*, 82–91.

Matsatsinis, N. F., & Siskos, Y. (2003). *Intelligent support systems for marketing decisions*. New York: Springer Science & Business Media.

Mirtalaie, M. A., Hussain, O. K., Chang, E., & Hussain, F. K. (2017). A decision support framework for identifying novel ideas in new product development from cross-domain analysis. *Information Systems, 69*, 59–80.

Montagna, F. (2011). Decision-aiding tools in innovative product development contexts. *Research in Engineering Design, 22*(2), 63–86.

Oh, J., Yang, J., & Lee, S. (2012). Managing uncertainty to improve decision-making in NPD portfolio management with a fuzzy expert system. *Expert Systems with Applications, 39*(10), 9868–9885.

Relich, M. (2013). Knowledge acquisition for new product development with the use of an ERP database. In *Federated Conference on Computer Science and Information Systems* (pp. 1285–1290).

Relich, M. (2015). A computational intelligence approach to predicting new product success. In *Proceedings of the 11th International Conference on Strategic Management and its Support by Information Systems* (pp. 142–150).

Relich, M. (2016). A knowledge-based system for new product portfolio selection. In P. Rozewski et al. (Eds.), *New frontiers in information and production systems modelling and analysis* (pp. 169–187). Cham: Springer.

Relich, M., & Pawlewski, P. (2016). A multi-agent framework for cost estimation of product design. In *International Conference on Practical Applications of Agents and Multi-Agent Systems* (pp. 73–84). Cham: Springer.

Shim, J. P., Warkentin, M., Courtney, J. F., Power, D. J., Sharda, R., & Carlsson, C. (2002). Past, present, and future of decision support technology. *Decision Support Systems, 33*(2), 111–126.

Sprague, R. H., & Carlson, E. D. (1982). *Building effective decision support systems*. New York: Prentice Hall.

Yang, Y., Fu, C., Chen, Y. W., Xu, D. L., & Yang, S. L. (2016). A belief rule based expert system for predicting consumer preference in new product development. *Knowledge-Based Systems, 94*, 105–113.

Chapter 5
Managerial Implications

This chapter aims to present the advantages and limitations of using the proposed approach in the context of the scenario and sensitivity analysis. These analyses are often based on the net present value (NPV) to evaluate economic efficiency of NPD projects.

Net Present Value Analysis

The NPV is an intuitive concept that assumes that one monetary unit today is worth more than one monetary unit in the future. An NPV analysis aims to evaluate the present value of future inflows and outflows. The NPV is the sum of discounted cash flows (CF) received in t time periods at the interest rate r:

$$\mathrm{NPV} = \sum_{t=0}^{n} \frac{\mathrm{CF}_t}{(1+r)^t} \tag{5.1}$$

The NPV illustrates the potential value that an NPD project adds to the company. A positive NPV indicates that the projected earnings generated by a new product exceed the anticipated costs. If there are many alternative projects, the project with the higher NPV should be selected. The interest rate (also called the discount rate or discount factor) refers to the company's opportunity cost of capital. This rate can be considered as the reward for accepting delayed payment and investing in the project rather than in other investments. The discount rate is often used to all investment decisions in companies.

M. Relich, *Decision Support for Product Development*, Computational Intelligence Methods and Applications,
https://doi.org/10.1007/978-3-030-43897-5_5

The Traditional Approach

The NPV analysis compares cash inflows (sales revenues) and outflows (costs) in the expected life cycle of a successful new product. Cash outflows include costs related to product development, marketing, and production. Costs of product development depend on the scope of market analysis, concept generation, product design, and prototype tests. Cash flows over the life cycle of a typical successful product are presented in Fig. 5.1.

The present value of a 6-year period is discounted at 12% per year (3% per quarter) back to the first quarter of the first year. For example, the present value for the fifth quarter is calculated as follows:

$$PV = 90{,}000 / (1.03)^5 = 77{,}635$$

Cumulative cash flows in the basic variant are illustrated in Fig. 5.2. Development cost is incurred before the launch of a new product. Marketing cost arises in the last phase of NPD, after launch, and then it supports sales. Production cost includes raw materials, components, and labour, and it is closely related to sales volume.

A sensitivity analysis aims to evaluate the impact of an individual parameter in an NPD model on the NPV through calculation of changes in the NPV corresponding to changes in parameters contained in the model. The obtained information helps the decision-maker identify areas for further improvements, e.g. cost reduction. Finally, the sensitivity analysis is useful for top management support in decisions relevant to concept selection and termination or continuation of an NPD project that is in progress.

Figure 5.1 illustrates a basic variant related to the expected sales revenue and costs that can be estimated on the basis of performance of a similar past product. The assumptions stated in the presented example contain the development budget of 1 million €, a 5-year period of cash flow analysis (the expected product life cycle), and production volume corresponding to sales volume. The basic variant supposes that development time lasts four quarters. To compare the impact of the development time on the NPV, the time was increased to five quarters (Fig. 5.3) and

	Year 1				Year 2				Year 3				Year 4				Year 5			
(in thousand)	Q1	Q2	Q3	Q4	Q1	Q2	Q3	Q4	Q1	Q2	Q3	Q4	Q1	Q2	Q3	Q4	Q1	Q2	Q3	Q4
Development cost	200	250	350	200																
Market analysis	130																			
Concept generation	70																			
Product design		200	270	150																
Prototype tests		50	80	50																
Marketing cost				100	100	120	120	120	100	100	100	80	80	80	50	50	50	30	30	30
Production cost					96	240	432	672	840	816	792	744	696	624	528	432	336	240	144	72
Production volume					200	500	900	1.400	1.750	1.700	1.650	1.550	1.450	1.300	1.100	900	700	500	300	150
Unit production cost					0.48	0.48	0.48	0.48	0.48	0.48	0.48	0.48	0.48	0.48	0.48	0.48	0.48	0.48	0.48	0.48
Sales revenue					180	450	810	1.260	1.575	1.530	1.485	1.395	1.305	1.170	990	810	630	450	270	135
Sales volume					200	500	900	1.400	1.750	1.700	1.650	1.550	1.450	1.300	1.100	900	700	500	300	150
Unit price					0.9	0.9	0.9	0.9	0.9	0.9	0.9	0.9	0.9	0.9	0.9	0.9	0.9	0.9	0.9	0.9
Cash flow	-200	-250	-350	-300	-16	90	258	468	635	614	593	571	529	466	412	328	244	180	96	33
PV, r = 12%	-200	-243	-330	-275	-14	78	216	381	501	471	441	413	371	317	272	211	152	109	56	19
Cumulative NPV	-200	-443	-773	-1.047	-1.061	-984	-768	-387	114	585	1.026	1.438	1.809	2.127	2.399	2.610	2.762	2.871	2.927	2.946
Total NPV of project	2.946																			

Fig. 5.1 The basic variant

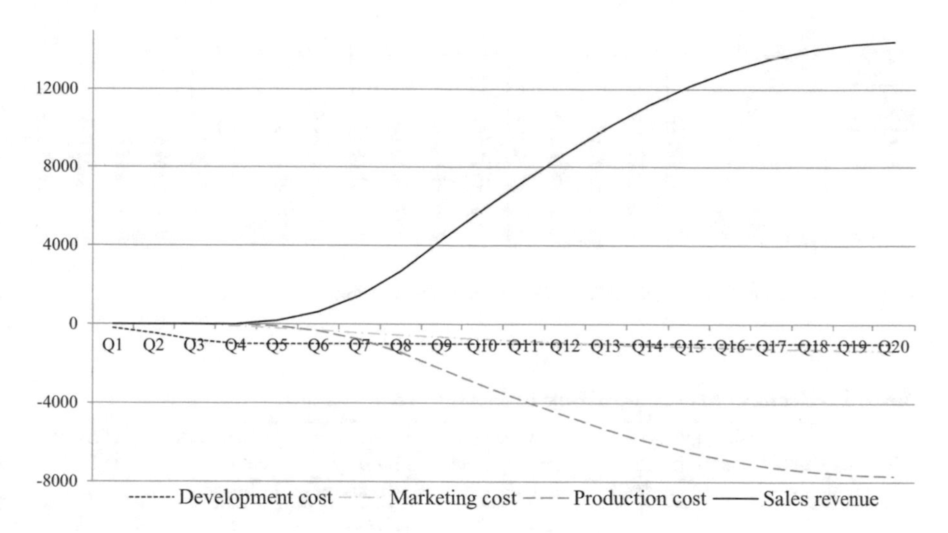

Fig. 5.2 Cumulative cash flows in the basic variant

	Year 1				Year 2				Year 3				Year 4				Year 5			
(in thousand)	Q1	Q2	Q3	Q4	Q1	Q2	Q3	Q4	Q1	Q2	Q3	Q4	Q1	Q2	Q3	Q4	Q1	Q2	Q3	Q4
Development cost	200	200	250	200	150															
Market analysis	130																			
Concept generation	70																			
Product design		160	200	150	110															
Prototype tests		40	50	50	40															
Marketing cost					100	100	120	120	120	100	100	100	80	80	80	50	50	50	30	30
Production cost						96	240	432	672	840	816	792	744	696	624	528	432	336	240	144
Production volume						200	500	900	1.400	1.750	1.700	1.650	1.550	1.450	1.300	1.100	900	700	500	300
Unit production cost						0.48	0.48	0.48	0.48	0.48	0.48	0.48	0.48	0.48	0.48	0.48	0.48	0.48	0.48	0.48
Sales revenue						180	450	810	1.260	1.575	1.530	1.485	1.395	1.305	1.170	990	810	630	450	270
Sales volume						200	500	900	1400	1750	1700	1650	1550	1450	1300	1100	900	700	500	300
Unit price						0.9	0.9	0.9	0.9	0.9	0.9	0.9	0.9	0.9	0.9	0.9	0.9	0.9	0.9	0.9
Cash flow	-200	-200	-250	-200	-250	-16	90	258	468	635	614	593	571	529	466	412	328	244	180	96
PV, r = 12%	-200	-194	-236	-183	-222	-14	75	210	369	487	457	428	400	360	308	264	204	148	106	55
Cumulative NPV	-200	-394	-630	-813	-1.035	-1.049	-973	-764	-394	92	549	978	1.378	1.738	2.047	2.311	2.515	2.663	2.769	2.824
Total NPV of project	**2.824**																			

Fig. 5.3 The variant with 25% increase in the development time

decreased to three quarters (Fig. 5.4). The development budget was proportionally divided between costs related to product design and prototype tests.

Table 5.1 presents changes in the development time and NPV in comparison with the basic variant. A 25% increase in the development time indicates a decline of the NPV from 2946 to 2824 thousands. In turn, a 25% decrease in the development time would increase the NPV to 3051 thousands.

Figures 5.5 and 5.6 illustrate changes in the NPV for a 10% increase and 10% decrease of the development cost, respectively. An increase of the development budget from 1000 to 1100 thousands leads to a decline of the NPV from 2946 to 2850 thousands. In turn, a 10% reduction of the development cost would increase the NPV to 3041 thousands.

Table 5.2 presents the NPV changes in the development cost. A 10% change in the development cost indicates a 3.2% change in the NPV.

The similar analysis conducted for other parameters of the NPD model leads to the comparison of affecting an individual parameter on the NPV. Table 5.3 presents

	Year 1				Year 2				Year 3				Year 4				Year 5			
(in thousand)	Q1	Q2	Q3	Q4	Q1	Q2	Q3	Q4	Q1	Q2	Q3	Q4	Q1	Q2	Q3	Q4	Q1	Q2	Q3	Q4
Development cost	200	300	500																	
Market analysis	130																			
Concept generation	70																			
Product design		230	390																	
Prototype tests		70	110																	
Marketing cost			100	100	120	120	120	100	100	100	80	80	80	50	50	50	30	30	30	30
Production cost				96	240	432	672	840	816	792	744	696	624	528	432	336	240	144	72	24
Production volume				200	500	900	1.400	1.750	1.700	1.650	1.550	1.450	1.300	1.100	900	700	500	300	150	50
Unit production cost				0.48	0.48	0.48	0.48	0.48	0.48	0.48	0.48	0.48	0.48	0.48	0.48	0.48	0.48	0.48	0.48	0.48
Sales revenue				180	450	810	1.260	1.575	1.530	1.485	1.395	1.305	1.170	990	810	630	450	270	135	45
Sales volume				200	500	900	1.400	1.750	1.700	1.650	1.550	1.450	1.300	1.100	900	700	500	300	150	50
Unit price				0.9	0.9	0.9	0.9	0.9	0.9	0.9	0.9	0.9	0.9	0.9	0.9	0.9	0.9	0.9	0.9	0.9
Cash flow	-200	-300	-600	-16	90	258	468	635	614	593	571	529	466	412	328	244	180	96	33	-9
PV, r = 12%	-200	-291	-566	-15	80	223	392	516	485	454	425	382	327	281	217	157	112	58	19	-5
Cumulative NPV	-200	-491	-1.057	-1.071	-991	-769	-377	139	624	1.078	1.503	1.886	2.212	2.493	2.710	2.866	2.979	3.037	3.056	3.051
Total NPV of project	**3.051**																			

Fig. 5.4 The variant with 25% decrease in the development time

Table 5.1 Sensitivities to changes in the development time

Development time (quarters)	Change in development time (quarters)	Change in development time (%)	NPV (thousands)	Change in NPV (thousands)	Change in NPV (%)
5	1	25	2824	−122	−4.2
4	0	0	2946	0	0
3	−1	−25	3051	105	3.6

	Year 1				Year 2				Year 3				Year 4				Year 5			
(in thousand)	Q1	Q2	Q3	Q4	Q1	Q2	Q3	Q4	Q1	Q2	Q3	Q4	Q1	Q2	Q3	Q4	Q1	Q2	Q3	Q4
Development cost	220	275	385	220																
Market analysis	140																			
Concept generation	80																			
Product design		220	297	165																
Prototype tests		55	88	55																
Marketing cost				100	100	120	120	120	100	100	100	80	80	80	50	50	50	30	30	30
Production cost					96	240	432	672	840	816	792	744	696	624	528	432	336	240	144	72
Production volume					200	500	900	1.400	1.750	1.700	1.650	1.550	1.450	1.300	1.100	900	700	500	300	150
Unit production cost					0.48	0.48	0.48	0.48	0.48	0.48	0.48	0.48	0.48	0.48	0.48	0.48	0.48	0.48	0.48	0.48
Sales revenue					180	450	810	1.260	1.575	1.530	1.485	1.395	1.305	1.170	990	810	630	450	270	135
Sales volume					200	500	900	1.400	1.750	1.700	1.650	1.550	1.450	1.300	1.100	900	700	500	300	150
Unit price					0.9	0.9	0.9	0.9	0.9	0.9	0.9	0.9	0.9	0.9	0.9	0.9	0.9	0.9	0.9	0.9
Cash flow	-220	-275	-385	-320	-16	90	258	468	635	614	593	571	529	466	412	328	244	180	96	33
PV, r = 12%	-220	-267	-363	293	-14	78	216	381	501	471	441	413	371	317	272	211	152	109	56	19
Cumulative NPV	-220	-487	-850	-1.143	-1.157	-1.079	-863	-483	19	489	930	1.343	1.714	2.031	2.304	2.514	2.666	2.775	2.831	2.850
Total NPV of project	**2.850**																			

Fig. 5.5 The variant with 10% increase in the development cost

	Year 1				Year 2				Year 3				Year 4				Year 5			
(in thousand)	Q1	Q2	Q3	Q4	Q1	Q2	Q3	Q4	Q1	Q2	Q3	Q4	Q1	Q2	Q3	Q4	Q1	Q2	Q3	Q4
Development cost	180	225	315	180																
Market analysis	117																			
Concept generation	63																			
Product design		180	243	135																
Prototype tests		45	72	45																
Marketing cost				100	100	120	120	120	100	100	100	80	80	80	50	50	50	30	30	30
Production cost					96	240	432	672	840	816	792	744	696	624	528	432	336	240	144	72
Production volume					200	500	900	1.400	1.750	1.700	1.650	1.550	1.450	1.300	1.100	900	700	500	300	150
Unit production cost					0.48	0.48	0.48	0.48	0.48	0.48	0.48	0.48	0.48	0.48	0.48	0.48	0.48	0.48	0.48	0.48
Sales revenue					180	450	810	1.260	1.575	1.530	1.485	1.395	1.305	1.170	990	810	630	450	270	135
Sales volume					200	500	900	1.400	1.750	1.700	1.650	1.550	1.450	1.300	1.100	900	700	500	300	150
Unit price					0.9	0.9	0.9	0.9	0.9	0.9	0.9	0.9	0.9	0.9	0.9	0.9	0.9	0.9	0.9	0.9
Cash flow	-180	-225	-315	-280	-16	90	258	468	635	614	593	571	529	466	412	328	244	180	96	33
PV, r = 12%	-180	-218	-297	-256	-14	78	216	381	501	471	441	413	371	317	272	211	152	109	56	19
Cumulative NPV	-180	-398	-695	-952	-966	-888	-672	-292	210	680	1.122	1.534	1.905	2.222	2.495	2.705	2.857	2.966	3.023	3.041
Total NPV of project	**3.041**																			

Fig. 5.6 The variant with 10% decrease in the development cost

Table 5.2 Sensitivities for changes in the development cost

Development cost (thousands)	Change in development cost (thousands)	Change in development cost (%)	NPV (thousands)	Change in NPV (thousands)	Change in NPV (%)
1100	100	10	2850	−96	−3.2
1000	0	0	2946	0	0
900	−100	−10	3041	96	3.2

the NPV for the optimistic and pessimistic variant in the context of changes in development cost, marketing cost, unit production cost, unit price, and sales volume.

The impact of an individual parameter of the NPD model on NPV significance can be concisely illustrated in a tornado chart (Fig. 5.7).

The Proposed Approach

The NPD model specified in terms of a declarative representation consists of a set of variables and constraints, including relationships between variables. As a result, the change of a single parameter of the NPD model may cause changes in other model parameters. For example, involvement of additional employees in product design and prototype tests may increase the reliability of a new product and, consequently, increase sales volume by reducing the cost of after-sales service. Then paradoxically, the increase of the development time and cost may result in increasing the NPV of an NPD project. Figure 5.8 illustrates a case in which the increment in the development time and costs (200 thousands in the fifth quarter) results in increasing the sales volume and, finally, the NPV of an investment.

Figure 5.9 presents a case in which the decrease in the development time results in reducing potential sales volume and the total NPV of an NPD project. The comparison of Figs. 5.8 and 5.9 with Figs. 5.3 and 5.4 shows that the change of the development time causes changes in sales volume and corresponding production volume.

The traditional approach makes available the NPV analysis through changes in parameters that directly affect the NPV (e.g. sales volume and costs related to product design, marketing, and production). In turn, the proposed approach enables simulations not only for main parameters of the NPD model but also for parameters that indirectly influence the NPV (e.g. the number of project team members who design and test a new product) and generate the cost of product design that finally affects the NPV. Information about optimal values of indirect variables seems to be more useful for the decision-maker taking into account the possibility of controlling an NPD project. The traditional approach provides information about desirable changes in costs and sales volume, but it is too general and abstract to implement desirable changes in project performance.

Table 5.3 Sensitivity analysis of model parameters

Changed parameter	Value for basic variant (thousands)	Optimistic variant (thousands)	Change (%)	NPV (thousands)	Change in NPV (%)	Pessimistic variant (thousands)	Change (%)	NPV (thousands)	Change in NPV (%)
Development cost	1000	900	−10	3041	3.2	1100	10	2850	−3.2
Marketing cost	1340	1206	−10	3049	3.5	1474	10	2843	−3.5
Unit production cost	0.48	0.43	−10	3509	19.1	0.53	10	2382	−19.1
Unit price	0.9	0.99	10	4003	35.9	0.81	−10	1889	−35.9
Sales volume	16,050	17,655	10	3439	16.7	14,445	−10	2453	−16.7

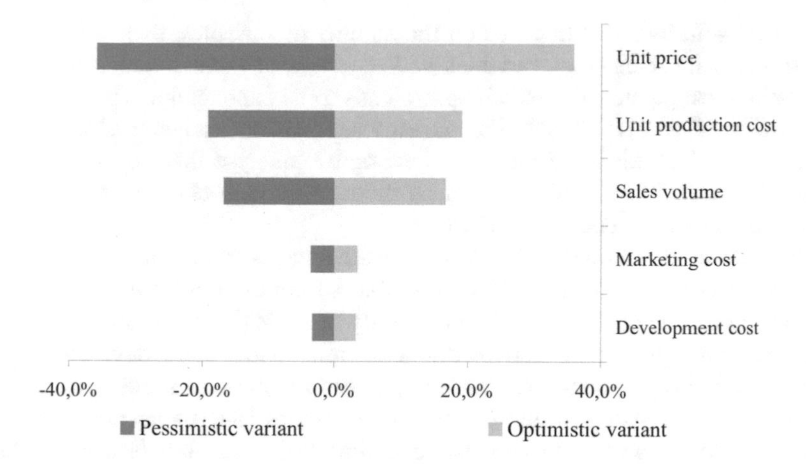

Fig. 5.7 Tornado chart for NPV changes

	Year 1				Year 2				Year 3				Year 4				Year 5			
(in thousand)	Q1	Q2	Q3	Q4	Q1	Q2	Q3	Q4	Q1	Q2	Q3	Q4	Q1	Q2	Q3	Q4	Q1	Q2	Q3	Q4
Development cost	200	250	350	200	200															
Market analysis	130																			
Concept generation	70																			
Product design		200	270	150	150															
Prototype tests		50	80	50	50															
Marketing cost					100	100	120	120	120	100	100	100	80	80	80	50	50	50	30	30
Production cost						104	262	475	746	941	922	903	856	807	730	623	514	403	290	176
Production volume						216	545	990	1.554	1.960	1.921	1.881	1.783	1.682	1.521	1.298	1.071	840	605	366
Unit production cost						0.48	0.48	0.48	0.48	0.48	0.48	0.48	0.48	0.48	0.48	0.48	0.48	0.48	0.48	0.48
Sales revenue						194	491	891	1.399	1.764	1.729	1.693	1.604	1.514	1.369	1.168	964	756	545	329
Sales volume						216	545	990	1.554	1.96	1.921	1.881	1.7825	1.682	1.521	1.298	1.071	840	605	366
Unit price						0.9	0.9	0.9	0.9	0.9	0.9	0.9	0.9	0.9	0.9	0.9	0.9	0.9	0.9	0.9
Cash flow	-200	-250	-350	-200	-300	-9	109	296	533	723	707	690	669	626	559	495	400	303	224	124
PV, r = 12%	-200	-243	-330	-183	-267	-8	91	241	421	554	526	498	469	427	369	318	249	183	132	71
Cumulative NPV	-200	-443	-773	-956	-1.222	-1.230	-1.139	-898	-478	76	602	1.101	1.570	1.996	2.366	2.684	2.933	3.116	3.248	3.318
Total NPV of project	**3.318**																			

Fig. 5.8 The variant with 25% increase in the development time

	Year 1				Year 2				Year 3				Year 4				Year 5			
(in thousand)	Q1	Q2	Q3	Q4	Q1	Q2	Q3	Q4	Q1	Q2	Q3	Q4	Q1	Q2	Q3	Q4	Q1	Q2	Q3	Q4
Development cost	200	300	500																	
Market analysis	130																			
Concept generation	70																			
Product design		230	390																	
Prototype tests		70	110																	
Marketing cost			100	100	120	120	120	100	100	100	80	80	80	50	50	50	30	30	30	30
Production cost				90	223	397	612	756	726	697	647	599	530	444	359	276	194	115	57	19
Production volume				188	465	828	1.274	1.575	1.513	1.452	1.349	1.247	1.105	924	747	574	405	240	119	39
Unit production cost				0.48	0.48	0.48	0.48	0.48	0.48	0.48	0.48	0.48	0.48	0.48	0.48	0.48	0.48	0.48	0.48	0.48
Sales revenue				169	419	745	1.147	1.418	1.362	1.307	1.214	1.122	995	832	672	517	365	216	107	35.1
Sales volume				188	465	828	1.274	1.575	1.513	1.452	1.349	1.247	1.105	924	747	574	405	240	119	39
Unit price				0.9	0.9	0.9	0.9	0.9	0.9	0.9	0.9	0.9	0.9	0.9	0.9	0.9	0.9	0.9	0.9	0.9
Cash flow	-200	-300	-600	-21	75	228	415	562	535	510	486	444	384	338	264	191	140	71	20	-14
PV, r = 12%	-200	-291	-566	-19	67	196	348	457	423	391	362	321	269	230	174	123	87	43	12	-8
Cumulative NPV	-200	-491	-1.057	-1.076	-1.009	-813	-465	-9	414	805	1.167	1.487	1.757	1.987	2.161	2.284	2.371	2.414	2.426	2.418
Total NPV of project	**2.418**																			

Fig. 5.9 The variant with 25% decrease in the development time

The identified cause-and-effect relationships between input and output variables of the NPD model and their usage in project management can be included in advantages of the proposed approach. However, the process of finding solutions of NPD-related problems also encounters some difficulties that can be seen as limitation of using the proposed approach. These difficulties are related to the number of admissible solutions that depends on the number of variables, their domains, and constraints. The number of variables and/or ranges of their domains involved in simulations are greater; the search space tends to be exponential. This increase in the number of admissible solutions is associated with an adverse effect on time needed to find solutions (if there are any) and to analyse them by the decision-maker. The mentioned difficulty can be reduced through changes of granularity related to domains of decision variables.

The presented example aims to illustrate advantages from reducing granularity in order to obtain a smaller set of possible solutions. Let us assume that the decision-maker wants to check the possibility of achieving the total NPV at 3300 thousands through changes in two parameters related to product development: the number of prototype tests (X_{15}) and employees involved in product design and tests (X_{11}). In the basic variant, the total NPV reaches 2946 thousands by the cost of product design and prototype tests equalling 800 thousands (see Fig. 5.1). In the basic variant, three employees are assigned to product design and its tests. The number of prototype tests was initially planned in 80 cycles. The decision-maker considers involvement of one additional designer to the project team, and the increase of the overall tests to 120 cycles. The extended period of product development causes the increase of costs reaching 200 thousands in the fifth quarter (see Fig. 5.8). Moreover, the decision-maker wants to simulate the changes in sales volume resulting from the extended period of time to product design and its tests and their potential impact on the reliability of a new product. Consequently, the NPD model should include, apart from the above-mentioned constraints, also cause-and-effect relationships between sales volume, the number of prototype tests, and employees involved in product design. These relationships can be identified with the use of data related to the performance of similar previous products.

An admissible change of a single parameter (X_{11} or X_{15}) does not fulfil the constraint referring to the NPV. Figure 5.10 illustrates changes in the NPV that depend on changes in both parameters (X_{11} and X_{15}). There are 53 possible solutions that fulfil all constraints. Numbers in square brackets represent the NPV, X_{11}, and X_{15}, respectively.

The number of possible solutions can be reduced through smaller granularity of a decision variable. Figure 5.11 illustrates possible solutions for the above-presented constraints and the number of cycles (X_{15}) divided by ten, namely from 8 to 12. There are six possible solutions that have their equivalents in solutions presented in Fig. 5.10. For example, the first solution denoted by 1# in Fig. 5.11 is equivalent to solution denoted by 6# in Fig. 5.10.

Another possibility for reducing the number of solutions presented for the decision-maker is the selection of only one solution that is optimal in the context of chosen criteria (e.g. the maximal NPV), or a few best solutions that can be easily considered in the decision-making process.

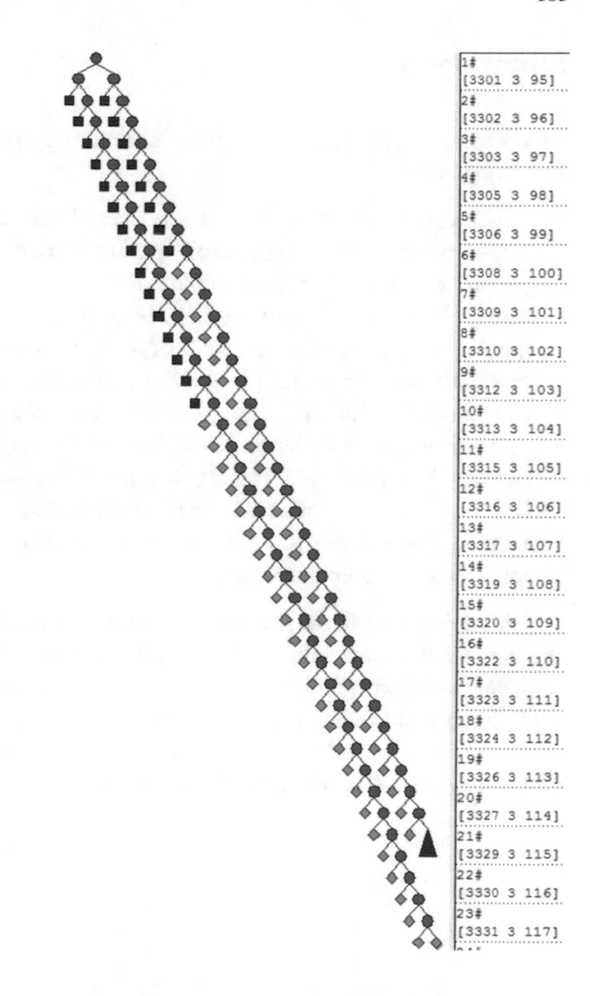

Fig. 5.10 Solutions for smaller granularity of X_{15}

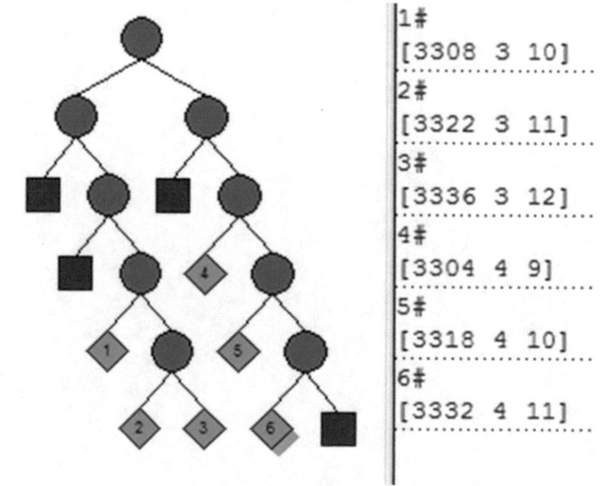

Fig. 5.11 Solutions for greater granularity of X_{15}

Summary

The proposed approach exceeds the traditional approach for the NPV analysis in the following areas:

- Parameters of the NPD model are interrelated (through cause-and-effect relationships identified with the use of parametric estimation), and the change of one parameter causes changes in others.
- The problem specified in the inverse form allows the decision-maker to verify quickly whether the desired NPV of an NPD project exists or not; if there is a set of solutions (variants to achieve the acceptable NPV), then the decision-maker can obtain all admissible combinations of chosen parameters of the NPD model.
- The presence of indirect variables in the NPD model enables implementation of changes in project performance towards achieving the desired NPV (the traditional approach provides information about desirable changes in costs and sales volume, but it seems to be too general and abstract to manage the NPD project towards a desirable outcome).

On the other hand, the proposed approach encounters some difficulties related to the number of admissible solutions that depends on the number of variables, their domains, and constraints that exist in the NPD model. The search space increases according to additional decision variables and/or ranges of their domains involved in simulations. This difficulty can be limited to some extent through changes of granularity related to domains of decision variables.

Chapter 6
Conclusions

Successful development of new products and their introduction into the market are key factors in improving business competitiveness. Nowadays, the majority of companies are using information systems to improve efficiency of business processes related to sales, marketing, manufacturing, purchasing, warehousing, accounting, and product design. Management information systems systematically increase their databases that may be a source of potential useful information for managers. This monograph is concerned with presenting a methodology for decision analysis and simulation modelling in order to support decision-makers in evaluating the potential of an NPD project and its possible implementation by desirable outcomes. The presented methodology may be useful for managers in project-oriented enterprises that develop modified versions of existing products.

The proposed methodology is intended to formulate a single NPD model within a declarative representation for specifying various NPD-related problems, e.g. portfolio selection of NPD projects, resource allocation, and evaluation of the project's potential. These problems are specified in terms of a constraints satisfaction problem that consists of variables, their domains, and constraints. As constraints refer to available resources (e.g. the number of employees, NPD budget) and relationships between variables, NPD-related problems may be stated in a forward and inverse form. Consequently, decision-makers may obtain information about the predicted value of an outcome (e.g. the profit from a new product, the cost of an NPD project) and about possible variants (if they exist) of alternative project completion (e.g. for reaching the desirable cost of an NPD project). Both forms of the stated problem use cause-and-effect relationships to prediction and simulations, respectively.

Current methods dedicated to evaluation of the potential of an NPD project use mainly an analogical reasoning and/or parametric estimation based on linear regression. Nevertheless, these methods have the limited ability to identify complex and imprecise relationships among the data related to the product development environment. To improve the results of parametric estimation regarding the project's potential, some computational intelligence techniques have been proposed. Artificial

M. Relich, *Decision Support for Product Development*, Computational Intelligence Methods and Applications,
https://doi.org/10.1007/978-3-030-43897-5_6

neural networks are used to identify relationships between the data specified in the precise form and then compared to multiple regression. In turn, neuro-fuzzy systems are applied to also include the impreciseness that may derive from the customers' opinions about a new product. The performed experiments show significant improvement in accuracy of identifying non-linear relationships using computational intelligence techniques compared to multiple regression. The use of computational intelligence requires data preprocessing, designing, and learning a neural (neuro-fuzzy) network, but more precise estimates of costs, sales volumes, and profits regarding new products support decision-makers in selecting NPD projects to a portfolio, allocating additional resources to a project, or withdrawing it from a project portfolio. Consequently, computational intelligence techniques may be considered as a pertinent tool for decision analysis in the context of NPD.

Computational intelligence techniques are able to identify cause-and-effect relationships that may be successfully used not only for prediction but also for simulations. The proposed approach allows the decision-maker to obtain information about the possibility of reaching the desirable project outcome. A declarative representation of NPD-related problems enables the identification of all possible solutions (if they exist), the number of which depends on constraints (including cause-and-effect relationships) and domains related to decision variables. As decision variables are related to project parameters (e.g. the number of team members, prototype tests), each solution may be considered as a variant for executing an NPD project in an alternative way. Consequently, the proposed approach is able to find a wider range of variants for economic effects of product development in comparison with the traditional scenario analysis that usually includes three scenarios: basic, optimistic, and pessimistic. On the other hand, the search space of possible solutions may be vast imposing the use of effective methods for reducing this space. The presented simulations show that constraint programming solves the CSP effectively, resulting in an improvement of interactive properties of a decision support system.

Specification of the NPD model in terms of the CSP enables the formulation of several NPD-related problems within a single NPD model. A declarative representation facilitates the NPD model updating towards adding variables and constraints related to diverse business areas and enables the specification of NPD-related problems in the forward and inverse form alternately. In a declarative approach, the main emphasis is put on the description of the desirable results without explaining the specific algorithms needed to achieve these results. If the NPD model includes relevant variables and coherent constraints, then the formulated problem will be solved. A declarative representation of the NPD model allows the decision-maker to obtain information about possible variants for reaching the desirable value of an outcome (e.g. the cost of an NPD project) and tracking difficulties in project performance. Simulations can also present the amount of resources that should be acquired to complete an NPD project. As a result, the decision-maker is supported towards solving the problem of project portfolio selection and resource allocation between NPD projects.

The presented methods of using computational intelligence for parametric estimation and constraint programming for simulations enable enhancement of decision analysis in the NPD process, particularly in the early phases related to project portfolio selection, product design, and prototype tests. Computational intelligence techniques are able to successfully identify relationships within multidimensional data structures that include variables referring to a product, company, and its environment that are often described in an imprecise form. These cause-and-effect relationships are further used to simulate the results of managerial decisions within project management and improve the NPD project performance. Nevertheless, the applicability of the proposed approach depends on some requirements within project management standards, including project planning and project performance measurement.

Glossary

Analogical reasoning Making predictions in terms of individual occurrences through involvement of past experiences to solve a problem that is similar to problems solved before. For example, an analogical approach may be used to search the most similar previous product(s) to a new product in order to identify a way of solving an existing problem, or to predict difficulties that can happen during new product development.

Artificial neural network (ANN) A network that is capable of modelling non-linear functions through creating connections between processing elements, i.e. artificial neurons. An ANN is able to identify relationships between input and output variables in so-called learning phase and predict the value of an output variable.

Computational intelligence (CI) A unified and comprehensive platform of conceptual and computing endeavours of fuzzy sets, neurocomputing, and evolutionary methods that have been inspired by observing biological systems to solve problems of information processing that are ineffective or unfeasible when solved with traditional approaches based on statistical modelling.

Computer-aided design (CAD) An information system that enables designers to design and manage their projects. CAD software is dedicated to create computer models of technical drawings.

Constraint programming (CP) An alternative approach to programming that embraces reasoning and computing, in which the main notion is a constraint on a set of variables that is a relation on their domains. CP consists of phases such as the specification of a problem in terms of constraints and a solution of this problem.

Constraint satisfaction problem (CSP) A problem defined as a finite set of variables, their domains, and a finite set of constraints that restricts the values of variables. A solution to a CSP is a set of the values of variables such that the values are in the domains of the corresponding variables, and all constraints are satisfied.

M. Relich, *Decision Support for Product Development*, Computational Intelligence Methods and Applications,
https://doi.org/10.1007/978-3-030-43897-5

Customer relationship management (CRM) A system dedicated to integrate marketing, sales, and after-sales activities in an enterprise. CRM systems include data about customers' history, including their sales volume, complaints, marketing efforts, and needs that can be used for concept generation related to a new product.

Decision analysis The process of formulating the decision problem, identifying cause-and-effect relationships, generating alternatives for solving this problem, assessing possible impacts of each alternative, evaluating alternatives, and selecting the best one according to preferences of decision-makers.

Decision support system (DSS) An interactive computer-based information system dedicated to support users in decision-making activities. DSS helps model decision problems (e.g. project portfolio selection, resource allocation, supplier selection) and identify the best alternatives.

Declarative representation Representation that is related to knowledge in a format that may be manipulated, decomposed, and analysed through a reasoning engine independently of its content, i.e. that expresses the logic of a computation without describing its control flow. As a result, the knowledge can be used in ways that could be omitted by the system designer.

Enterprise resource planning (ERP) An information system dedicated to registration, planning, and performance monitoring of some business processes. An ERP system provides an integrated view of core business processes across various departments (e.g. accounting, purchasing, manufacturing, sales) using a common database.

Fuzzy logic An approach based on fuzzy sets that may represent non-statistical uncertainty and approximate reasoning through using a form of many-valued logic, in which the concept of partial truth is employed. As a result, fuzzy logic is much closer to human reasoning and natural languages that are usually imprecise.

Knowledge base A collection of facts and rules for describing a particular domain of knowledge and written in a knowledge representation language, e.g. predicate calculus. It can be seen as a database of rules about a subject considered in knowledge-based systems.

Knowledge management The process of capturing, cleaning, storing, distributing, and using the knowledge and information according to business goals.

Management information system (MIS) A computer-based system that assists in the performance of management functions in an enterprise. A MIS can be used to analyse data and visualise information in order to support users in their business activities, including planning, controlling, and decision-making.

Net present value (NPV) A method for calculating the present value of present and future cash flows of an investment. The NPV can be used to compare the potential of new products according to their expected cash flows and life cycles.

Neuro-fuzzy system (NFS) A fuzzy system that is trained by a learning algorithm derived from neurocomputing. This hybrid system combines advantages of fuzzy systems and an ANN, being a universal approximator with the ability to interpret if-then rules.

New product development (NPD) The process of introducing a new product into the market. NPD consists of a series of phases, including market analysis, concept generation, product design, prototype tests, manufacturing, and commercialisation.

Parametric estimation A method for predicting an output variable (e.g. project cost, time) using input variables that significantly influence an output variable. A parametric model includes an analytical function that describes relationships between input and output variables.

Product life cycle The length of time between product conceptualisation and product removal from the market. The main stages of this cycle embrace: research and development, introduction into the market, growth, maturity, and decline in sales.

Project management The process of initiating, planning, executing, monitoring, and closing a set of activities related to a project within specific objectives (e.g. time, cost, quality).

Scenario analysis The process of estimating various states of a system and generating strategies to try to identify potential management challenges. Scenario analysis is often used to evaluate the expected value of an investment or business activity, towards analysing the results of decisions, e.g. related to project portfolio selection.

Scheduling The process of arranging work in an optimal way (e.g. towards minimising the production time and costs). In the context of NPD project portfolio, scheduling is used to allocate resources (e.g. employees, machines, financial means) effectively between NPD projects.

Sensitivity analysis A tool to evaluate the results of specific changes in values of input variables impacting on an output variable. For example, an increase of the NPD cost by 1% reduces the NPV of 2%.

Simulation modelling The process of a computer simulation that is based on a model of the behaviour of a system, and aims to produce the outcomes of interest. Simulation modelling can generate alternatives for decision-making and examine potential consequences of decisions.

Index

M. Relich, *Decision Support for Product Development*, Computational Intelligence Methods and Applications,
https://doi.org/10.1007/978-3-030-43897-5

Printed in the United States
by Baker & Taylor Publisher Services